Handmade Brick for Texas

Handmade Brick for Texas

A Mexican Border Industry, Its Workers, and Its Business

Scott Cook

LEXINGTON BOOKS
A division of
ROWMAN & LITTLEFIELD PUBLISHERS, INC.
Lanham • Boulder • New York • Toronto • Plymouth, UK

Published by Lexington Books
A division of Rowman & Littlefield Publishers, Inc.
A wholly owned subsidiary of The Rowman & Littlefield Publishing Group, Inc.
4501 Forbes Boulevard, Suite 200, Lanham, Maryland 20706
http://www.lexingtonbooks.com

Estover Road, Plymouth PL6 7PY, United Kingdom

British Library Cataloguing in Publication Information Available

Library of Congress Cataloging-in-Publication Data
Cook, Scott (Howard Scott), 1937–
Handmade Brick for Texas : A Mexican Border Industry, Its Workers, and Its Business / Scott Cook.
pages cm
Includes bibliographical references and index.
ISBN 978-0-7391-4797-9 (cloth : alk. paper) — ISBN 978-0-7391-4798-6 (pbk. : alk. paper)
1. Brickmaking—Mexico, North. 2. Brick trade—Texas. I. Title.
TP826.5.M6C65 2010
338.4'766673709721—dc22 2010041558

™ The paper used in this publication meets the minimum requirements of American National Standard for Information Sciences—Permanence of Paper for Printed Library Materials, ANSI/NISO Z39.48-1992.

Printed in the United States of America

Mexican Brick

Ladrillo tras ladrillo
Mexicano
Ladrillo tras ladrillo
Hecho a mano
Labor-intensive handmade
Mexican brick
Both sides of the Rio Grande
Smoky brickyard villages
De puros mexicanos
Men, women, children
Rising from the delta mud
Artisans of river clay
People of the ladrilleras
Manos hechas callo
Spirits hardened far
Beyond the calloused palms
Beyond the calloused knees
In ritual beside a
Mound of muddy clay
One thousand bricks
Per man per day
Ladrillo tras ladrillo
Mexicano
Ladrillo tras ladrillo
Hecho a mano
Mexican brick

Rosa Canales Perez

March 2005

Contents

Maps

Tables

Photographs

Preface

This book is a sequel to *Mexican Brick Culture in the Building of Texas, 1800 – 1980s* published in 1998. More detailed and focused in scope and content than my earlier book, it concentrates on the handmade brick industry in Reynosa, Tamaulipas, Mexico, tells the stories of its *ladrilleras* [brickworks or plants] and *ladrilleros* [brickmakers] in greater depth than was possible in the earlier book, and presents an updated view of the industry's competitive situation in the Texas market.[1]

The stories come from the memory banks of informants who were either still working in or retired from the brick industry at the time I interviewed them. Since the industry suffered a major loss of market share during the 1980s that greatly reduced the number of operating plants, many informants I interviewed in the 1990s could only speak about their past experiences in the industry. For them this meant the decades of the 1950s, 1960s, and 1970s, when the industry flourished to supply the booming speculative building market in Texas. Nevertheless, the lower Rio Grande/Rio Bravo corridor's brick industry does not yet fully reside in the museum of the extinct, and its living representatives are not merely witnesses for a livelihood lost. Handmade brickmaking is ongoing in the Reynosa area today and there are many brickmakers who make a living from it. Consequently, discussions about brickmaking by retired brickmakers remain pertinent to the craft as it is still practiced.

Much material presented in this book was not used in the earlier book and derives from more than fifteen months of research conducted in the lower border region from 1987 through 1995. Between 2000 and 2008, I conducted additional research in Reynosa, some of which has provided the basis for publications (Cook 2001; Cook 2005; Cook 2006a and b). In 2006, I visited the Acme Brick plant near Elgin, Texas to gather data about its now defunct 'soft–mud' operation, and in 2007 did the same at the new Reynoso family plant in Reynosa. This most recent material is presented in chapter 9.

I wrote previously that "between 1960 and 1985 several billion Mexican handmade bricks, worth hundreds of millions of dollars, were exported to Texas" (1998:140). This estimate was based upon available statistics, including those from a survey conducted by the *Brick Institute of Texas*, and was probably inflated. The truth is that we will never know exactly how many Mexican handmade bricks, manufactured in the riverine zone of the state of Tamaulipas from Nuevo Laredo to Matamoros, were exported to the U.S. during that twenty–five year period. We can, however, rest assured that the volume was substantial, certainly in the hundreds of millions, and that its value amounted to many tens of millions of dollars. It can also be asserted with confidence that the volume from 1985 to the present is much reduced from previous levels. There are only a handful of Reynosa brick plants in business today compared to many dozens in the twentieth century boom period. By and large, these surviving plants are more dedicated to the industry, better managed and equipped, and more responsive to market fluctuations and nuances.

The basic theoretical concerns underlying my interest in the handmade brick industry are essentially the same ones that have stimulated my research in Mexico since the 1960s, namely, how artisans and workers engaged in various kinds of commodity production make and earn a living, how the enterprises in which they participate confront changing environmental, social and economic conditions, and, most importantly, how these various subjects relate to, are impacted by, and are engaged with the evolving division of labor and market structure in North American and global capitalism. I addressed these issues in *Understanding Commodity Cultures,* and examined Mexican brick culture at various points in my exploration of them (Cook 2004:334 for index listings).

One important theme that I addressed in my 1998 book is the relationship between agriculture and brickmaking in the lower Rio Grande/Rio Bravo region. The role of large–scale, irrigated agriculture in the Rio Grande Valley is well–known and, indeed, is part and parcel of the twentieth century history of urbanization and economic growth of deep south Texas. Less well–known is the role of agriculture on the Tamaulipas side of the region, especially beginning in the late 1930s. A recent pioneering study by Casey Walsh (2008) focuses on the twentieth–century development of irrigated cotton agriculture in the Tamaulipan portion of the lower Rio Bravo delta. Walsh's study has important implications for understanding the interplay between different levels and sectors, public and private, within Mexican capitalism, as well as the role of the U.S. government and capital, in the formation of the region's political economy and political culture. Starting in the late 1930s, the Mexican federal government, arranged for the repatriation of Mexican agricultural workers in the United States to colonize and participate in the establishment of newly formed irrigation districts for cotton cultivation in areas immediately to the west [*Distrito de Riego del Rio San Juan*] and east [*Distrito de Riego del Bajo Rio Bravo*] of Reynosa. Agrarian reform measures associated with this project made relations between *colonos*, *ejidatarios*, and *pequeños propietarios* part of an emerging regional class formation. This clearly had implications for the growth of the regional brick industry by acting as a magnet for labor recruitment and investment capital. The contradictions at play in the national, regional, and local discourses revolving around development and maldevelopment that Walsh (2008:167–169) analyses so astutely in the case of the emergence of Valle Hermoso as a separate municipality with a "new geopolitical identity," reverberate with discourses expressed among brickmakers in the periurban *colonias*/*ejidos* of Reynosa. Indeed, as the reader will learn in several chapters of the present book, many of these brickmakers are themselves *colonos*, *ejidatarios*, or *pequeños propietarios*, or had experiences with such persons either in agriculture or brickmaking.

Turning to another issue, data on costs, prices, wages, and profits are plentiful and important in any economic study. Nevertheless, in studies of the border economy, their analysis can be especially problematic because the calculation of monetary values is complicated by the fact that business is transacted in two currencies, the U.S. dollar and the Mexican peso. In this dual currency economy, peso to dollar rates of exchange are critical to the process of determining values. It is up to the analyst to make certain that economic value data are as accurate as possible. I have tried to do this by cross–checking sources, for example, comparing estimates from different informants, by comparing informant estimates with my own observations during appropriate time periods, and by comparing informant estimates with official and unofficial data in published sources.

The most reliable reference point I can provide for 1970s values derives from Terra–Mex plant records that included copies of union contracts (Cook 1998: Ch.9). In 1970 the contract stipulated a rate of 27.50 pesos per 1000 standard size bricks molded; the 1974 contract stipulated a rate of 49.50. The peso to dollar exchange rate in 1970 was 12.50 and averaged about 17.00 in 1974. This means that a molder received only about $2.20 [U.S.] per 1000 bricks in 1970 and $2.90 [U.S.] in 1974. For the 1990s the benchmark data comes from *Ladrillera Reynosa, S.A.* plant records for November, 1992 when 27.280 pesos was the piecerate per thousand bricks [8–inch length] molded. The peso to dollar exchange rate at that time was 3.11, so the piecerate in dollar terms was $8.77. The benchmark pricing data for 2007 comes directly from plant records kindly provided by *Ladrillera Reynoso* and are analyzed in chapter 9.

Once the problem of accuracy has been addressed and the most reliable data presented, the real complications regarding their interpretation begin—especially when time perspective is critical. It is axiomatic in economics that monetary values change through time so that, for example, the value of a dollar in 1935 was different from its value in 1945, 1955, 1965, 1975 and so on. This is true whether its value is defined in terms of purchasing power or in terms of its exchange rate for other currencies like the Mexican peso. Since the U.S.–Mexico border economy operates in terms of two currencies, people and businesses operate on a quotidian basis in both currencies. Given its participation in a dual currency economy, the border brick industry has evolved as a dual currency industry in ways that will be described and analyzed in several chapters of this book.

One practical consequence of the dual currency system is that participants in the border brick industry have to think about and try to remember relevant figures—prices, costs, wages, and profits—in both currencies because their economic experiences have been mediated by and/or represented in both dollars and pesos. Workers, historically speaking, have earned their livelihoods in dollars on the U.S. side of the border and in pesos on the Mexican side and, as consumers, have also spent on goods and services in both currencies. Employers have mostly sold their bricks priced in dollars but have met their costs of production in pesos—a transactional bifurcation that has made business more profitable, if also more tricky in terms of accounting.

The Mexican peso, more than the U.S. dollar, tends to complicate analysis for, at least, two reasons. First, it is the currency of a national economy that has suffered more severe inflationary crises over the last several decades than its U.S. counterpart. Second, the Mexican government has found it necessary to intervene radically in its monetary system in recent decades to address problems of overvaluation/undervaluation of its currency vis–à–vis the dollar. One need only mention the decade between 1982–83 and 1992–93 characterized by several such interventions culminating in the introduction in January 1, 1993 of the New Mexican Peso equivalent to 1,000 former Mexican Pesos. This was a challenging period for everyone involved with the border economy, as evidenced by the proliferation of '*casas de cambio*' [currency exchange businesses] surrounding border crossing points. The dollar–to–peso exchange rate went from 12.50 in 1955 to 3,121 at the end of 1992, with two intervening jumps from 23 in 1981 to 150 in 1982 and from 922 in 1986 to 2,260 in 1987.

Exchange rate fluctuations in a dual currency economy complicate calculations of all agents across the board. So, the task of getting the record straight and of interpreting monetary values in such an economy is especially challenging. Accordingly, I have made a systematic effort throughout the text to convert pesos to dollars at the exchange rate prevailing in the year corresponding to the data. Thanks to the internet, there are easily accessible sources presenting these exchange rates in time series.[2]

In my interviews with them, informants nearly always provided estimates of piece rates, wages and approximate dates of employment when discussing their work histories—some, admittedly, with less certitude than others. Most of them also expressed a general sense of the declining purchasing power of wages over the course of their careers, some of which spanned several decades during which both money wages and prices increased from single to multiple digits. Nevertheless, there clearly have been fluctuations over the years in real wages in the border brick industry, always complicated by fluctuations in peso/dollar exchange rates. This makes it difficult to generalize about workers' standard of living in real terms during the second half of the twentieth century.

In 1988, regarding the Mexican economy as a whole, I wrote (Cook 1988:57): "Looking at these trends in time perspective from a 1977 base line, and taking as a measuring rod the value of the peso in 1977 (a 1977 peso + 100 centavos), the following scenario obtains: the peso has steadily declined in value since 1977 reaching fifty–seven centavos in 1980, fourteen centavos in 1983 and only three centavos in 1986. Whereas the average minimum daily wage has risen steadily in money terms from eighty–eight pesos in 1977 to 137 pesos in 1980 to 429 pesos in 1983, and to 2,046 pesos in 1986, it has done so . . . at a rate which has not kept pace with the rate of inflation and peso devaluation. Taking into consideration both movements, it can be calculated that, whereas the real minimum daily wage was eighty–eight pesos in 1977, it had declined to a paltry fifty–eight pesos in 1986, or only 66.7 percent of the 1977 minimum wage in terms of real purchasing power. Considered from a 1983 base line, the real purchasing power of the minimum wage in mid–1987 had declined to only fifty–two percent of that of the 1983 minimum wage. Finally, given the continuing devaluation of the peso [which has been allowed to freely float against the dollar] imported goods have become too expensive for the shrinking incomes of most Mexicans. All of this translates into a progressively severe reduction in the standard of living for the working class."

In retrospect, the conditions outlined in this statement not only provide a rationale for the phenomenon of undocumented immigration from Mexico to the United States during the period in question, and subsequently, but also support a general sense among informants in my study that income earned from work in the brick industry has progressively declined in terms of the standard of living it can buy. It must be noted, however, that the regime of Carlos Salinas de Gortari between 1988 and 1994, through a series of market–oriented reforms, was able to reverse some of the trends outlined in my 1988 article. For example, a 160 percent rate of annual inflation in 1987 was reduced to single–digit levels in 1993 and average real wages in manufacturing rose some 30 percent between 1988 and 1992, an average annual rate of 6.8 percent. The upshot of these data is that major economic indicators shift in Mexico from *sexenio* [six year presidential term] to *sexenio* due to economic policy changes, as well as in response to global market conditions.

Photography has been an integral part of my research on the brick industry. Over the last fifteen years I have taken several hundreds of photos of brick plants and of all stages of brickmaking. At appropriate places in every chapter I have made a systematic effort to refer the reader to photos supplementing the text. Since several figures and tables are inserted within chapters, I decided to place photos in a separate appendix (see Appendix 1). These photos are consecutively numbered and their specific page location and captions in Appendix 1 are listed in the table of contents. I urge the reader to supplement his or her reading of the text with viewing the appropriate photos. Brickmaking, like all craft industries, is best approached through a combination of textual description and visual images.

In December, 2005 I was privileged to have the opportunity to participate in a program on 'Border Brick Culture' sponsored by the city government of Reynosa. The program included a performance by Joe and Rosa Pérez of songs popular among the brickmakers and other working people in the borderlands that was preceded by my overview of brick culture in Reynosa and the lower Rio Grande border. The program was well received by those attending and I was especially impressed by the degree to which its content resonated with, and intersected the biographies of, several city officials who were present. One of them disclosed that his grandfather had been among the city's pioneering leaders of the brick industry whose name I had mentioned in my presentation. Another told us that he had owned a neighborhood business patronized over the years by many brickmakers and had personally experienced improvised evening musical sessions involving workers from a nearby brick plant—and including most prominently the now legendary artists of *ranchera* music, Cornelio Reyna and Ramón Ayala. This experience has reinforced my belief that brick culture runs deep in the historical memory and public consciousness of Reynosenses today. It is my sincere hope that this monograph resonates positively with these currents in Reynosa life, and stimulates wider popular interest in the municipality's unique role for over a century as a host of the regionally important brick industry.

The connection of aesthetics to the handmade brick industry does not end with the incidental role of music in the lives of brickmakers. Indeed, the very persistence of demand for authentic handmade brick in a twenty–first century Texas building market dominated by large–scale mechanized industry reflects aesthetics more than economics. No one has explained why this is so more convincingly than A.B. Searle (1956:115): "hand–moulding is by no means an obsolete method, for by it bricks of a beautiful appearance can be made better than by many other method. It is costly and slow, and skilled moulders are scarce, so that the use of machinery has many advantages; but where a beautiful appearance is an important consideration, hand–made bricks still occupy the first place." This is as applicable to England in the last century as it was to Texas then and today.

Notes

1. I used the English term 'brickyard' as a translation of the Spanish term '*ladrillera*' in my two previous books (Cook 1984 and Cook 1998). Incidentally, in Spanish '*ladrillera*' is a simplified version of the more cumbersome term '*ladrillería*' and is more widely used is most Spanish–speaking countries today. The only use of the latter term I have encountered during my research was in Camargo where the now–defunct Camargo Brick Company was identified in Spanish as *Ladrillerías Camargo*.

In the present book, I use either the English term 'brickworks' or the compound term 'brick plant' as a translation for *ladrillera*. This change reflects my preference to accept the Spanish meaning of *ladrillera* as the physical place where brick are made, as distinct from *almacén* or *negocio de venta de ladrillos* for the place where finished brick are stored for sale. This latter meaning I reserve for the term 'brickyard' in English.

2. Two useful internet websites for dollar–to–peso exchange rates are: 1) *Relacion Historica Tipo de Cambio Dolar/Peso. Taller de Actualización de la Información Financiera. Normas del Boletín "B–10" Reconocimiento de los Efectos de la Inflacción en la Información Financiera*; 2) A series of three spreadsheets listed as pesos and dollars.xls containing data compiled by *El Centro de Estudios de las Finanzas Públicas de la H. Camara de Diputados* from the Banco de Mexico and the Federal Reserve Bank of St. Louis. The three spreadsheets cover the period from 1980 to 2007. One spreadsheet focuses on *Sobre o Subvaluación del Peso*; another at *México:Tipo de Cambio Nominal FIX*; and the third at *México: Tipo de Cambio Real*. These are downloadable files available at the following website: www.cefp.gob.mx/intr/estadisticas/esta37.xls

Acknowledgments

I gratefully acknowledge the cooperation and assistance of all of those persons, agencies, offices, and organizations listed in my 1998 book *Mexican Brick Culture in the Building of Texas, 1800s–1980s* (Texas A & M University Press). This includes many persons identified and quoted in the present monograph. Without listing their names again here, I reiterate my gratitude for their patience and cooperation with my questioning. The late Ray Rohlfs, of what was then U.S. Brick and is now Hanson Brick, generously supported my project during the summer of 1993, and was also an interested and knowledgeable informant. Major funding for the research was provided by National Science Foundation grant SBR–9404609, with supplements from the University of Connecticut's Research Foundation and Institute for Puerto Rican and Latino Studies.

I also take this opportunity to acknowledge the following persons who have been especially cooperative and helpful in my continuing work on the border brick industry: Rolando García and Carmen Pérez have played a key role in introducing me and my project to the architectural and building community through the Lower Rio Grande Valley chapter of the American Institute of Architecture (LRGV–AIA). Thanks to Rolando's familiarity with my 1998 book, he and Carmen arranged for my collaboration with Joe and Rosa Pérez [*Rumbo al Anacua*], music educators and performers from Olmito, through our participation in the programs of the 2005 and 2006 Building Communities Conferences at the South Padre Island Convention Center, as well as in the program of the 2006 convention of the Texas Society of Architecture in Dallas.

Rosa's reading of my 1998 book inspired her to write a poem entitled "Mexican Brick" [*Ladrillo Mexicano*] reprinted here as the Epigraph (p. iv). She and Joe collaborated to perform songs mentioned in my 1998 book, together with other songs in the border ranchera/corrido tradition, in a program "Songs of the Border Brick Culture" that was presented in the above–mentioned events and several times subsequently.

George and Virginia Gause of the Library at The University of Texas Pan American in Edinburg have also been cooperative and supportive of my work over the years, as has Adela Ortega, Director of the Mission Historical Museum. Also, Raciel Juárez has shown interest in my research and was instrumental in organizing a presentation of the "Songs of the Border Brick Culture" for officials of the City of Reynosa in December 2005, as well as a program on "Border Brickmakers" for Los Caminos del Rio at the Museum of South Texas History in Edinburg on June 12, 2008. Shan Rankin and her staff also co–sponsored that program, and have been supportive of my work on this topic over the years.

James Nader, past President of the Texas Society of Architects, was instrumental in supporting the 'Songs of the Border Brick Culture' program held at the 2006 convention of the Texas Society of Architecture in Dallas. He also invited me to accompany him on an informative tour of Acme Brick's plant in Elgin, Texas where we witnessed some of the final runs of their 'soft–mud' operation. I also thank Jim Krueger, manager of that operation, for his guided tour on February 24, 2004, and to D. Leon Prince, general manager of Acme's Elgin plant, for his patience in answering my follow–up questions and for authorizing me to take photographs during my solo return visit to the plant on February 27, 2004.

Ann Cook, photo librarian for the Texas Department of Transportation in Austin, was very generous with her time and knowledge in retrieving, from the archive she so ably manages, photographs taken by TXDOT photographers in the early 1950s of the brick plant in Relámpago. I am grateful for her help in obtaining copies of these unique photographs. Staff librarians at The Center for American History at The University of Texas in Austin were also helpful in providing me with information and an orientation regarding The South Texas Border: The Robert Runyon Photograph Collection.

I also wish to thank Doug Boyd of Prewitt & Associates in Austin, whose work as a salvage archaeologist has put him in the trenches with the remains of many old brick plants and brick buildings in the Rio Grande Valley and elsewhere in Texas, for his careful reading of a draft version of this book. His comments and suggestions were very helpful to the process of correction and revision.

I would be remiss, in closing, to fail to mention the hospitality and cooperation for many years of several representatives of the Reynosa brick industry. I am especially indebted to two Antonios, "Toño" Reynoso and his son Tony, to Joel Rodríguez Flores, and to doña Minerva Morales de Garcia without whose help my project would have been much more difficult and much less rewarding. Finally, thanks to the magical power of serendipity, during 2008 I met and interviewed two important past players in the border brick business, namely, Charles [Chuck] Revelette of San Antonio and Lester Tisdale of Pharr. I gratefully acknowledge cooperation from them in sharing with me the content of their memory banks regarding their careers in the border brick business.

Chapter One
Introduction

For a variety of reasons, the state of Texas provides a unique laboratory for looking at persistence and change in clay brick production and the use of clay brick in construction. Foremost among these is its sheer size and demographic growth from 1900 to the present, combined with a tradition of brick production and brick construction dating back to pre–statehood days of the nineteenth century. As in other areas of the United States, the growth of mass markets in Texas necessarily transformed brick production: labor–intensive, artisanal technology and small–scale familial organization was replaced by capital–intensive, mechanized technology, and large–scale corporate organization. The sheer volume of brick required to meet the demands of Texas' expanding construction market could be produced only by opening more plants, and/or by building larger–scale, more highly mechanized plants.

Nevertheless, this process has been influenced by a persisting aesthetic demand for artisanal brick that, in turn, has affected brick production and marketing. Hence, the juxtaposed development of two kinds of mechanized technologies of brickmaking known as soft–mud and stiff–mud, terms that differentiate between the higher clay and lower water composition of the stiffer material compared to the softer material. Soft–mud processes are designed to simulate handmade brick, whereas stiff–mud processes are designed to mass produce in the most technologically advanced, efficient, and least costly way brick that are highly uniform in size, shape, and texture. Brick produced by these two processes do have different attributes that can be objectively described but their overall appearance architecturally [i.e., as face brick on buildings] is definitely a matter of subjective taste (see Gurcke 1987: 13–23).

Another factor that makes Texas a unique laboratory for examining the brick industry is its historical relationship with and proximity to Mexico. The lower Rio Grande/Rio Bravo border area provides a special illustrative case. After 1900, as a by–product of the massive hydraulic agricultural and settlement/urbanization project evoked by slogans like "Magic Valley" and "Gringo Builders" (Allhands 1931), brickmaking flourished in the lower border area of the Rio Grande Valley. Anglo–Texan and Tejano capital employed mostly Tejano and undocumented Mexican labor to establish brick plants, both semi–mechanized and unmechanized, in Cameron, Hidalgo, and Starr counties. Brickmaking was also undertaken at or near construction sites with provisional scove–type kilns or clamps (Campbell and Pryce 2003:306).

There was a surge in regional aesthetic demand for antique brick in Texas during the late 1950s and early 1960s. Since the mechanized brick factories in Texas were relatively unsuccessful in producing brick with an antique or handmade look to capture this growing market, the door of the Texas market was opened for Mexican handmade brick. Ironically, economic factors, especially availability and relatively low price, and not the aesthetic factor, assured the flow of these brick into the Texas mass market during the 1970s (Steinbomer 1981:46–47), predictably engendering a competitive response by Texas producers.

The booming Mexican handmade brick business during those decades was quite in harmony with the persisting aesthetic appeal of artisanal brick among Texas architects. This was exemplified most notably by O'Neil Ford both rhetorically and through his building projects. It is with regard to the complexities of the twentieth century record of border brickmaking, because of its involvement with the off–border Texas masonry construction market and with competitive forces impacting Texas brick producers, that the architectural role of handmade brick becomes at once most relevant, ironic, and problematic.

On the one hand, architects professionally participate in the celebration of historic buildings built with handmade brick and in their restoration. On the other, some of them either promoted or collaborated in the proliferating use of Mexican artisanal brick during the building boom of the 1960s and 1970s. Although the same aesthetic factor can be seen as a rationale for both types of participation, there is little doubt that in the speculative building market of the 1960s and 1970s the high demand for Mexican handmade brick was more a function of the temporarily short supply of U.S. machine–made brick, and of the plentiful supply of relatively low–priced Mexican brick, and not of its aesthetic appeal. This is not to discount the appeal of the handmade brick 'look' among building professionals and home buyers.

The opinion that "old brick and brick made in Mexico were aesthetically more pleasing" than U.S. machine–made brick, shared by O'Neil Ford and other Texas architects, is subjective and, for some, arbitrary. The particular appeal of artisanal brick, for Ford and others with his architectural tastes, resides in a combination of factors such as natural earthiness and softer appearance that makes these brick indispensable for a building that "has humility and suitability to its site, surroundings, function, and historical context." According to his San Antonio associate, Allison Peery, Ford's architectural practice reflected his aesthetic judgment: "He [Ford] would dilute our pristine designs with his damned old Mexican brick . . . [making the] work unique and significant and warm and human and all the good things it was and is" (George 1992:90; cf. pp. 105, 173).

Implicit in Ford's doctrine and illustrated in practice during his career was a connection of the aesthetic factor to brick production—its location, technology, and organization. Ford himself, and his brother Lynn, periodically traveled to the border or to the interior of Mexico to find handmade brick suited to particular projects. During the 1950s, reportedly to assure a supply of "brick with character," Ford initiated a relationship with artisanal producers in Piedras Negras, Coahuila across the river across from Eagle Pass. His choice of low–tech producers can be interpreted to mean that the type or style of brick that most appealed to him could not be produced to his satisfaction by high–tech processes (George 1992:90).

Ford's practice of what he preached apparently had enough of an impact on the Texas brick market in the 1950s to motivate Acme Brick Company to pioneer the production of a machine–made Mexican replica line known as *El Ford*, specifically designed and produced for a large Ford project in Oklahoma. According to Luke Gresham, engineer–in–chief of Acme plants at the time [interview with author 5/19/93]: "He [Ford] wanted a Mexican brick so we went ahead and made him an ASTM machine–made brick . . . that fell into that category aesthetically." George (1992:90) comments on Acme's experiment as follows: "The big brick manufacturers . . . became concerned about lost business and sent letters to members of the architectural profession warning that salvage brick and Mexican brick were technically inferior. They also began to offer more attractive choices: the Acme Brick Company introduced the 'El Ford' as their response to Ford's challenge to 'match this brick'" [i.e., handmade that he was having produced in Piedras Negras].[1]

As a "modernist" Ford was certainly no enemy of machine technology as shown by his promotion of the "Lift–Slab" method of construction (Dillon 1999:20). Yet his association with Acme's 'El Ford' experiment was arguably at odds with his preference for low–tech artisanal brick, as was the use of Acme's machine–made brick in the Trinity University project. Despite widespread plaudits for and the undeniable attractiveness of the Trinity buildings, there is some evidence in the conversational record to suggest that Ford would have preferred to see "damned old Mexican brick" on the walls of those buildings.

There is also evidence, however, that Ford was receptive to brick produced through appropriately engineered technology incorporating varying degrees of mechanization. First, his Skidmore College project involved the development of a special brick composed of a mixture of burned and over–burned clay to reproduce the color and texture of brick produced in the region at the beginning of the twentieth century (George 1992:160). Second, he enthusiastically endorsed the ill–fated effort at the small–scale yet appropriately mechanized, state–of–the–art Nordmeyer plant in Rio Grande City to produce a rough, irregular brick with all the texture and flavor of old–style brick but with none of its weaknesses. Nordmeyer brick was intended not to be confused with the typical Mexican brick but, instead, to reproduce the aesthetic qualities of colonial handmade brick as used to rebuild Williamsburg, Charleston, and other historical building sites (see Cook 1998: Ch.7). With the mark NSW, this brick was specifically engineered to withstand severe weathering conditions in Texas.

Unfortunately, the appropriate engineering prowess of the Nordmeyer plant to produce a high–spec handmolded brick was never put to test in the Texas brick construction market due to its premature closure. So, O'Neil Ford's reported enthusiasm about Nordmeyer brick, probably as much a reflection of his "passion for place" (Dillon 1999:1) as of his sense of aesthetics, was economically inconsequential.

A main underlying theme in the present study revolves around this intriguing intersection of aesthetics, technology, and economics in the Texas brick industry. This has been expressed through repeated attempts over a period of several decades by the largest Texas–owned and –based brick company, Acme Brick, to successfully engineer a product to supply a persisting demand for a rustic, rough, irregularly shaped, adobe–looking brick in the residential and commercial building markets where architectural design and style are paramount. O'Neil Ford and Acme Brick, therefore, appropriately become background protagonists in my narrative about Mexican handmade brick.[2]

As will be explained in chapter 9, Acme Brick's recent closure of the soft–mud production line at its plant in Elgin, Texas, left Mexican handmade plants from Piedras Negras to Reynosa without a Texas competitor in the Fordian market niche. So, a competitive process initiated over a half–century ago to mass–produce brick with a handmade look through state of the art machinofacture has come full circle in the first decade of the twenty–first century. Perhaps this is not an event of world–historic importance but it is uniquely demonstrative of the techno–economic pitfalls that bedevil the successful engineering of the mass production of simulated handmade brick.[3]

The Riverscape and Mexican Border Brick Culture

The term "riverscape" was coined by Paul Horgan (1984) in his classic history of the Rio Grande to convey the idea that landscape created by it should not be viewed as static. In his words: " . . . the river has been surrounded by forces and elements constantly moving and dynamic, interacting to produce life and character" (1984:6–7). He goes on to argue that while the "main physical characteristics of the Rio Grande seem timeless and impersonal" they "assume meaning only in terms of people who came to the river" (1984:7). Horgan's words ring true for me after more than two decades of fascination with the deltaic region of the lower Rio Grande corridor and especially with one category of people who have lived and worked there for generations, the *ladrilleros* or brickmakers. Although on different time scales, the movement and dynamism of the meandering river that has deposited clays and sand is matched by that of the brickmakers whose livelihood depends upon these raw materials.

Without the riverscape's water and alluvium, Mexican brick culture could not have developed industrially in the lower Rio Grande border region. Brickmaking shares agriculture's dependency upon the riverscape's twin resources but is unique among non–agricultural manufacturing industries both for its pioneering regional industrial status, and for its unique extractive and transformative use of these resources. It was among the region's earliest industries that extracted and transformed raw materials into commodities on a commercial basis.[4]

The special nature of the deltaic Rio Grande's physical environment emerges from a careful study of Map 1 that shows the riverine brickmaking belt of Reynosa, Tamaulipas and Hidalgo County, Texas. The most important brickmaking sites are indicated by black dots, and populated places, including Reynosa's *colonias*, are indicated in white letters. The magnetic north symbol is located in a white box in the map's lower left hand corner; the scale of the map [upper right hand corner] in its original size was 2.25 inches = 1 mile but in reduced size becomes roughly .9 inches = 1 mile. The Rio Grande/Rio Bravo itself is indicated in white letters and the international boundary, that follows the river, is indicated by a broken white line. [See Appendix 1. Photos 8, 9 and 10 provide views of the river as an international boundary from three different lower border locations].[5]

Tracing the broken white line indicating the river boundary from left to right across Map 2 gives the reader a bird's eye view of the incredibly meandering path of the river channel as it courses its way through the brickmaking belt. Not to be mistaken for the river in examining this map is the comparatively straight black line running from left to right across its lower portion. This black line represents the Anzaldúas Diversion Canal [*Canal de Derivación Anzaldúas*], one of Mexico's main irrigation canals that begins eight miles upriver from Reynosa at Anzaldúas dam and runs through the city to the agricultural fields in the Lower Rio Bravo Irrigation District [*Distrito de Riego Del Bajo Rio Bravo*] lying midway between Reynosa and Matamoros, and extending southwards beyond Valle Hermoso to encompass a significant chunk of northeastern Tamaulipas (Walsh 2008).[6]

The juxtaposition of the meandering river and the unswerving canal symbolizes the natural and man–made dimensions of this unique deltaic riverscape as do bridges, dams, levees [*bordos*], pumping stations, networks of drainage and irrigation ditches, settling basins [*esteros*], ox–bow lakes [*resacas*], wells [*norias*], and *bancos* and/or *rincones* [landforms between meanders in the river channel]. Wherever brickmakers have practiced their craft in this region some, if not all, of these features are located nearby, thus lending a special character to the industrial setting.[7]

The brickmaking process is more complicated than it may first appear to the casual observer. The process involves muscle power and physical endurance as well as generous inputs of applied science and artisanry. The applied scientific domain involves such factors as knowledge of clay properties and climatic conditions, site development and organization, and the technology or procedures that determine how brick are to be molded, dried, and fired so as to meet accepted standards. The lack of uniformity in this array of factors translates into a certain degree of regionalization, and even localization, of Mexican brick culture. This means that a period of orientation is required before even the most experienced brickmakers from, say, Oaxaca, Guanajuato, San Luís Potosí, Coahuila, Durango, Veracruz or Nuevo León can contribute optimally to brickmaking in the Lower Rio Grande region. For example, brickmolders from Oaxaca who use open frame molds have to learn how to use the box mold with a closed bottom. Consequently, owing to sub–regional variations in clay properties and climate, brickmaking in Matamoros will have subtle, nuanced differences from brickmaking in other parts of riverine Tamaulipas [see Map 1], and more pronounced differences from brickmaking in other states.

Mexican brick culture is transmitted between generations from masters to apprentices, a process that in the interior and on the border typically, though not exclusively, occurs through patrilineal kin relations: father to son, grandfather to grandson, uncle to nephew, and sometimes including females as in father to daughter, uncle to niece, and so on. Brick culture is practiced in *ladrilleras* [brick plants or brickworks] of different sizes and with different degrees of mechanization of the manufacturing process. In the brick plants which are the focus of this study, clay extraction, preparation, and transport may be partially or completely mechanized but the clay–molding process is done exclusively by hand thus justifying the definition of the brick so molded as handmade. Typically, throughout Mexico, once dried, brick are set in updraft kilns that are used repetitively for many batches but are fired and cooled–down for each batch; firing occurs at the bottom and heat and gases circulate upward (cf. Campbell and Pryce 2003:15, 308–309; Gurcke 1987:28–34).

In the lower border region artisanal brickmaking has been practiced primarily by Mexican–born brickmakers for at least 150 years. It continues to be practiced in the twenty–first century on the Mexican side of the border, albeit on a reduced scale, but has not been commercially practiced on the Texas side since the 1960s. The size or scale of artisanal brickmaking enterprises in the Lower Border during the second half of the twentieth century ranged from one or two workers to one hundred or more. The varying degrees of mechanization ranged from none [i.e., simple hand tools such as picks, shovels, hoes, and wheelbarrows] to the use of various machines such as mechanical diggers or backhoes, pug mills or clay mixers, forklifts, dump trucks, and flatbed tractor–trailers. In other words, mechanization may be found at the stages of clay extraction and preparation, firing, and transport in artisanal plants. The typical kiln in such plants is updraft, either wood– or gas–fired, but there is significant variation in the type of construction and firing capacity [ranging from 10,000 to 250,000 or more brick].

The following recollection by Ed Nordmeyer of the construction of the *Casa de Palmas* hotel in McAllen, Texas around 1918 suggests how brickmakers were respected by other workers for their combination of artisanry and capacity for hard labor:

> My first personal recollection of the brick industry anywhere in the Valley was in McAllen. The brick that went into the *Casa de Palmas* hotel in McAllen was

> made over on 16th Street, west of the present location of the hotel. The brick industry then was more of a cottage industry rather than a mechanized industry that we had in later years. The brick was made right there on the ground from the local clay. It was molded by artisans who were not *peons* although they were dressed in short pants, no shirt, and were working barefooted. They had large bladed hoes that they used to mix their mud with. They mixed it right there. They cast the brick into molds, turned the molds over on relatively dry, level ground, and then cleaned the mold, sanded it, and filled it again. Now these were folks normally addressed by others as *maestros* or masters. They were people that were looked up to because the way they made brick was halfway an art.

The term *maestro* in this context is a term of respect for a master of his craft. Nordmeyer also observed that the relatively small scove kilns at temporary brickmaking sites like this were made of unfired brick and mud, and were fired with green mesquite (see Gurcke 1987:29–32).[8]

To underscore the idea that skilled labor and knowledge underlie what appear to the casual observer to be simple, straightforward tasks like drying handmade brick, Ino Weiske, a Georgia Tech–educated ceramic engineer and son of Guenther Weiske of Valley Brick and Tile Company in Madero, Texas [see below], stated:

> People seem to think that the matter of drying handmade brick is very simple. It's not really because when they come out of that mold, you can't handle them. They are still too soft. And you want to get them turned on edge as soon as possible. It takes experience. It reminds me of when my doctor assured me that from an EKG he could tell in five minutes what's wrong with my heart. Now that is experience, and that is what it takes for handmade brick. You have to know when you can turn it on edge because as long as it's flat it doesn't dry well. You don't have enough surface exposed to the air. So, it's important to turn it on edge. Then, when they dry out a little bit more, you still have to get the right time, you stack it in a tower so that it gets more wind. So, it is not all that simple to dry handmade brick. I think drying is the most difficult part of the process.

Weiske may be right on this point but I am sure he would also agree that handmade brickmaking is a sequenced ensemble of challenging tasks starting with clay acquisition and preparation, proceeding through molding and drying, and culminating in firing and cooling down, all of which must be informed by the expertise of a master artisan to yield high quality brick.

Unquestionably the most interesting, skilled, and physically demanding of the jobs performed on any given day at handmade brick plants is that of *hechurero* [lit. one who makes or maker]. It embraces several specific tasks including preparing clay, molding, and demolding brick. Indeed, one of these tasks—the *cutting* [*cortar*] by hand of a glob of moist clay from a mound, usually to fill two compartments of a six–compartment box mold, itself gives rise to an alternative term for a brickmaker, namely, *cortador* or cutter.

The molding process involves individual molders in an almost choreographed sequence of coordinated movements of hands, arms, legs, and feet in an unbroken interactive process with adjacent supplies of clay, sand, water, and various tools or implements including the all–important wooden mold [*molde*]. After assuming a kneeling position, the always barefoot molder sands his mold from a nearby pile of coarse sand, sands his hands from a separate pile of even finer sand, then *cuts* a glob [*bolita*] of moist clay from the parent mound [*montón de soquete or barro batido*], enough to fill one or two compartments, depending upon the size of the mold, and gently rolls it in the fine sand before tossing and hand–packing it into the sanded mold. The sand in the bottom of the mold and on the hands of the molder minimizes clay adhesion. Typically sand is hand–sifted with a sifter [*criva*] to assure its uniformity. Molds have two, four or six compartments according to the size of the brick to be molded—the larger the brick the fewer the compartments.[9]

The molder repeats the procedure until all compartments of the mold are filled with moist clay, and then strikes off the excess clay with a wooden striker [*rascera*] that he quickly passes across the top of the mold. This entire process is completed in about one minute. At that point the molder stands up, picks up the filled mold and, in a slow trot, carries it to an appropriate place on the nearby drying floor, and turns it over [*voltear* or *volteada*] to gently deposit or demold [*desmoldar*] the moist brick in symmetrical straight rows on the smoothed–out dirt floor [*piso*] of the patio.

In turning over the mold the molder bends over from the waist down with arms fully extended, hands grasping the upper sides of the mold, and rests the mold ever so briefly as he carefully rotates it forward until the top of the mud–filled compartments lie completely on the ground, and the still moist brick release from the mold. Viewed from the front, the molder's head is at the apex of an isosceles triangle, his outstretched arms form the sides, and the upturned mold with its contents, the base. This action and its posture, comparable to that required in callisthenic bending and stretching exercises, is an impressive culmination to the brick molding sequence. The two sequenced actions, mold turnover and demolding have to be performed in a carefully controlled way to avoid twisting or otherwise warping the moist blocks of clay as they are released onto the drying floor.

Once brick are laid out on the floor to begin the drying process, the molder immediately returns to his clay mound, again assumes a kneeling position, and repeats the process. The same sequence is repeated over and over again, at a steady pace, until a targeted number of brick is molded and laid out to dry. Before that process is completed, at intervals of each twenty or so fillings and demoldings, the molder will have to dip the mold into a tank [*pila*] of water to wash off clay that has inevitably stuck to the corners and crevices of the mold.

The wetting of the mold also helps it to retain the sand required to minimize clay adhesion during the demolding process. The molder must also smooth out the surface of the sanded ground drying floor using a long–handled rake with a metal scraper on one end [*rastrillo*] to minimize marks or dents on the underside of demolded brick.

The typical box mold with metal *placas* [*frogs* or brand name plate indentation] for a common brick measures 30 x 10 x 3 inches, has six compartments each measuring 8–1/2 x 3–3/4 x 3 inches, an open top and closed bottom, and weighs about ten to fifteen pounds empty. Filled with wet clay the weight more than triples to between thirty–five and forty–five pounds. So, the repetitive athletic movements described above are performed by a molder carrying loaded molds on each first leg of the round trip from clay supply to deposit point on the molding floor, and empty molds on the return leg. It is understandable why I have rarely met an overweight brick molder, regardless of age; typically they are agile, wiry and have the physical endurance of experienced athletes. At the end of the day a molder has jogged as far as a long–distance runner but has done so carrying substantial extra weight.

The output of each day's molding is typically laid out in three or more columns or lines of six brick each conforming exactly to the pattern of the mold, with a space of about two inches between each column. Each column has at least fifty rows or 300 brick; the first row is about ten feet beyond the clay supply which means that the fiftieth row is some sixty feet from the molder's clay supply. This means that a daily output of 1,000 brick requires the molder to traverse a considerable amount of yardage, usually at a pace best described as a slow jog. Whether daily output is 1,000 or more brick, one of the most impressive sights at plant like this one is to see the results of a day's labor on display in columns and rows so symmetrically patterned as to suggest a computer–driven, professionally engineered production process rather than one involving a lone artisan guided only by his own vision and kinetic energy. [See Appendix 1. Photos 1–7 illustrate the steps and activities in the molding process described above].

If *ladrilleros*, and molders in particular, have special attributes as workers, Mexican border *ladrilleras* are also unique as work sites. This is especially true of the plants located in *ejidos* or semi–rural *colonias* in the urban periphery where land is also used for agriculture, animal husbandry, housing, or trash dumping/recycling. Although networks of unpaved roads and paths connect these different land use zones, fences and gates restrict entry to farms and ranches to keep livestock contained. Most brick plants, however, are open, located next to roads, and are easily accessible by pedestrians and vehicles. They typically have residences for workers but if worker households keep animals like chickens or pigs these are kept in pens or corrals to restrain them from roaming into the *patio* areas where they can trample upon newly molded brick. Stray or free roaming dogs are also relatively absent for the same reason—

making *ladrilleras,* for that reason alone, unique spaces in Mexico's *colonias populares* and *pueblos*.

It is ironic that *ladrilleras* organized to produce handmade brick that uniquely combine economic and symbolic values, themselves embody a unique human geography of production with its own architecture and appearance. The *ladrillera* landscape is gouged with clay pits and cluttered with tools and other objects but, also, encapsulates architecturally significant constructions like walled, arched and chambered kilns, as well as massive elongated, roofed molding and drying sheds. When viewed together with row–upon–row and stack–after–stack of symmetrically aligned brick, this totality of elements makes a strong visual impression. The geography of Rio Grande border brickmaking becomes even richer and more complex when interlaced with the equally distinctive riverine environment.

The Origins of *Maquila* in the Handmade Brick Industry

It is a little–known fact of lower Rio Grande border economic history that the origin of *maquila* industry in Reynosa precedes the National Border Program or *Programa Nacional Fronterizo* [PRONAF] and the *Programa de Industrialización Fronteriza* or the Border Industrialization Program [BIP], and lies decades earlier in the handmade brick industry of the late 1930s. This original *maquila* arrangement came about in the course of doing business, and was an improvised response to market conditions for brick and brickmaking labor. It occurred under the radar of government or private associational initiatives but, still, had significant economic consequences for participating enterprises.

Moreover, the unique nature of brickmaking gave its pioneering *maquila* involvement a more pristine form, closer to the original meaning of the term, than the forms that emerged later through the promotional efforts of the BIP. Expressed somewhat differently, the twin–plant arrangement in the handmade brick industry is closer in substance to the original meaning of the term *maquila* than is the type of shared production launched by the BIP.

In feudal Spain, and colonial Mexico, *maquila* referred to the portion of grain, flour or oil [in the case of olive pressing] charged by the miller or presser for his services, or to the lord of the manor or *hacendado* when appropriate. The verb *maquilar* referred to the process of measuring and charging this fee in kind. The commodities involved in this original *maquila* relationship were transformed without the addition of other component parts or materials—from wheat to flour, from corn as *nixtamal* [soaked corn] to *masa* [dough], from raw olives to olive oil. Likewise, in the 1930s–1960s border brick industry case, crude [green or unfired] clay that was hand molded and dried in Mexico was vitrified [chemically fused and hardened] by kiln–firing in a Texas plant through a *maquila* arrangement, thus being transformed from adobe–like blocks to real brick.

As the meaning of *maquila* evolved and was adapted to the conditions of modern mechanized industry, it came to refer to "any particular activity in an industrial process, such as assembly or packaging effected by a party other than the original manufacturer" (Hansen 2003:1). In the context of the industrial process set in motion by PRONAF/BIP, *maquilar* has assumed the specific meaning "to assemble in a cross–border plant", and *maquiladora* has been introduced to refer to a cross–border assembly plant (Oxford Spanish Dictionary, 1994 edition). As a leading scholar of the subject has expressed it (Hansen 2003:1): "modern *maquiladoras* constitute assembly operations for products which, after processing, are re–exported to the U.S. and other countries." Since assembly was never the issue in the border brick industry, the crucial factor was linked production in which commodities produced in Mexico in an unfinished state were exported to the U.S. for finishing and marketing.

The *maquiladora* or twin–plant program was viewed as a combined government/private sector response to conditions of unemployment in the border region owing to the termination in 1964 of the agricultural guest worker or *bracero* program and the repatriation of tens of thousands of Mexican migrant workers (Hansen 2003). Unlike agriculture, the brick industry never had a guest worker program, and underwent a reorganization and relocation as a consequence of the U.S. government's forced repatriation of undocumented Mexican workers in the 1950s.

The earliest record of systematic business relations of a *maquila*–type in the lower border brick industry date from the late 1930s and involved Guenther Weiske's Valley Brick and Tile Company [VBTC] in Madero, Texas that produced hand molded, as well as a variety of machine–made brick. Although Weiske's involvement in brickmaking dates back to 1911–12 when he produced hand molded brick at the *El Gavilán* ranch plant [near the river south of Madero on the road to Granjeno] that were used to build the nearby Oblate Novitiate, by 1923 he had relocated to a new plant in Madero that featured the production of machine–extruded brick. According to Guenther's son Ino, his father "was connected with and knew about handmade brick but figured that mechanically made brick was better and that was the way to go." In other words, Guenther Weiske's perspective in 1923 was that "machine–made brick was the bright thing of the future" (author's interview with Ino Weiske October 10, 1995).

This strategy brought the company success. Nevertheless, there remained a demand for handmade brick because it was cheaper and because it had an attractive appearance, unmatched by machine–made brick, so in the 1930s Weiske's plant renewed hand molded production, not because he had a particular fondness for the product but as a practical business decision motivated by the philosophy, as expressed by his son, of "if you can't beat 'em, join 'em." The plant's production of hand molded brick was dependent upon hiring hand molders from Mexico, since local workers had neither the skills nor the desire to perform this physically taxing and muddy work. Given the unavailability of guest worker programs for brickmaking, the hand molders employed by VBTC were undocumented [hiring of undocumented Mexican workers was not, at the time, a punishable offense]. Ino, still a high school student at the time, said that "in the late 1930s and early 1940s I can still remember just about biweekly visits from the Border Patrol that would kind of disrupt production." VBT was caught in a bind: they had a market for handmade brick but its production required the employment of undocumented workers and, therefore, periodic harassment by the Border Patrol.

To resolve this dilemma, Weiske went across the river to seek production arrangements with small Reynosa plants. According to Ino (5/25/93), "My father would go over there and contract with one or other small outfits along the river to the west of the bridge to supply green or crude brick" (ibid.). He related an anecdote involving his father at that time:

> He was over there [Reynosa] and asked around about people who had brick plants, and one man told him 'Oh yes, he knew so–and–so' and gave his name. Father asked him directions on how to get there, and this fellow told him, 'Well, you go two blocks down here and then you turn right and go two more blocks and turn left and then you come to a road that's heavily traveled and turn right, and you go about two kilometers out, and there's the plant.' Father said: 'Yes, but after all of those turns I might miss that heavily traveled road.' And the other fellow replied: 'No, the road is so heavily traveled that you can't miss it.' My father then asked him: 'Well, how come it's so heavily traveled?' And the fellow responded: 'It's the one that goes out to 'Boystown'' [local term for a red light district].

Ino Weiske summed up his recollection of this period succinctly: "We got into the business of getting brick made in Mexico, and we would ship them over here and fire them in our kilns." Another rationale for this *maquila* arrangement had to do with taking advantage of the gas–firing kiln technology at the Madero plant. Again to quote Ino: "I believe that at least a part of the reason for firing raw brick over here was that the sales people did not want to get sued for selling halfway raw brick. The thing is that part of the fad of handmade brick was pink color. But pinks are underfired and sometimes they got too underfired." Controlled firing in the kilns at the VBTC plant eliminated underfiring, but to still produce the desired pinkish tint, red sand or volcanic ash known as pozzolan, mined on VBTC land in Rio Grande City, was trucked to the Madero plant and to handmade plants in Reynosa to be used to sand molds and the patio drying floors. This meant, according to Ino, that "you could fire them pretty hard but they still had a little bit of pink color where they were sand–struck."

This business continued at least into the mid–1950s but was also supplemented by the importation of handmade brick that had been gas–fired in Reynosa at the Salinas plant. Given the high demand for handmade in those years, when possible, VBTC also produced it in Madero with undocumented Mexican labor (Cook 1998:108–111).

The story of the border brick industry's role in the prehistory of *maquila* started with VBTC but did not exclusively involve that enterprise. Several others were involved as well, most notably, Anderson's two–plant business divided between Olmito in Cameron County and Relámpago in Hidalgo County [see Map 1]. Anderson got into the brick business in 1946 by selling brick made on his large Olmito citrus farm for worker housing by his foreman, Alfonso Flores, to an interested client. One thing led to another and by 1947 Anderson got hooked on commercial brickmaking, relying heavily on his foreman Flores who was the son of a brickmaker from the state of Tlaxcala in Mexico, and who had prior experience in the craft both in Matamoros and in the Ruenes plant in Olmito, before coming to work for Anderson as farm supervisor. By 1947, near a *resaca* [oxbow lake] just east of old highway 77, kilns had been built, fired with an ample supply of mesquite from land cleared to plant oranges, patios and a drying shed constructed, and clay was being extracted from what would become deep pits. The kilns were converted to gas by 1948. This plant was so successful that Anderson soon acquired another one, Har–Lan Clay Products owned by C.L. Landrum and W.F. Hardwick, that was already up and running with gas–fired kilns in Relámpago near Progreso. The Relámpago plant was expanded in the early 1950s and became Anderson's main plant (Cook 1998: Chapter 5) [see ABY on Map 3].

At both of these plants Anderson experienced problems with his largely undocumented Mexican labor force owing to periodic raids by the Border Patrol that seem to have been triggered by pressure on the INS from organized labor, most notably the Teamsters Union due to its lack of success in attempts to organize the brickmakers. As early as 1948 Alfonso Flores, Jr. [b. 1932], who never followed in his dad's footsteps as a brickmaker ["I didn't like it, muddy all the time"], drove flatbed bobtail trucks, and later larger trailers, for Anderson, regularly hauling unfired or green brick from Matamoros to Olmito for firing there. Anderson leased a plant in Puertos Verdes on the outskirts of Matamoros, that was supervised by Alfonso Flores, Sr. on visits every day or two, strictly for the production of green brick to be fired in the gas–firing kilns at the Olmito plant. Green brick were also regularly purchased at other plants in Matamoros, most notably one owned by the Mejias family. Flores, Jr. even recalls hauling crude brick, for a period of two to three months, from as far away as San Pedro de Roma [renamed Ciudad Miguel Alemán in 1950] to Olmito for firing. During the five–year period that Flores, Jr. drove trucks for Anderson's Olmito plant he remembers hauling only unfired brick from Mexico back to Olmito for firing.

The same pattern of *maquila* production occurred at Anderson's Relámpago plant from the early 1950s until its shutdown in 1956–57. Through his plant foreman, Manuel Herrera, Anderson arranged access to land or fields [*campos*] on the Mexican side of the border for crude brick production under the supervision of Herrera or his Relámpago patio boss, Jesús Vargas. Luciano González, who worked at the Relámpago plant from 1948 to 1952–53, remembered that by the time he stopped working there "they were still burning brick but it became harder to do so because of the migratory labor problem. You had to operate the plant with fewer workers. The Border Patrol did not allow us to work" [*Todavía quemaban ladrillo pero entonces ya se puso duro por cuestión de la migración. Ya no se pudo trabajar con mucha gente. La migra no nos dejaban trabajar*—interview with author 5/3/95].

According to González, in response to the crackdown on undocumented workers, Anderson and Herrera started making crude brick on the Mexican side of the river in fields they had leased at various places near Progreso, Las Flores, and La Plancha. In his words: "They were making unfired brick [in Mexico], trucking them to Relámpago, and firing them there" [*Se hacían crudo y lo traían para Relámpago en camiones. Y los cocían en Relámpago*]. Sebastián Reynoso, who returned to Reynosa about the same time González did after several years of work at the Relámpago plant, continued to produce crude brick for that plant under its *Aztec* brand at his newly opened plant in *Ejido Los Longoria*. This was a contractual arrangement involving payment in advance for a stipulated quantity of brick, and typified the Relámpago plant's *modus operandi* with Mexican producers.

Prior to the completion of the international bridge between Progreso, Texas and Nuevo Progreso, Tamaulipas in December 1953, crude brick shipments would have had to be routed across the Reynosa–Hidalgo bridge. This meant that the earliest phase of maquila production involving the Relámpago plant was based in Reynosa through producers like Sebastián Reynoso who had previously worked in Relámpago prior to his deportation and repatriation. Soon after the Progreso bridge opening, however, crude brick were trucked to the plant from the Mexican side through Nuevo Progreso.

William C. Cain, whose father built and owned the Progreso bridge, opened up a customs brokerage to handle commercial shipments across the bridge, recalled that there was a regular movement of crude brick in 1954–55 on trucks belonging to the Relámpago plant. He did business only with Manuel Herrera, the plant supervisor. In his words (interview with author 4/26/95):

> I made a blanket carrier's certificate or an inward manifest carrier's certificate for the informal entries. And we just kept tabs on how many brick came over, and I'd make one entry per day. I think probably that he would bring at most two truckloads a day. He'd have one ready to go over when the bridge opened and then he'd have another one, because they close the customhouse at 5 p.m. for informal entries. He'd go over to Relámpago, unload the truck, get back over to Mexico, load it again and be back by 4 p.m., and take it to the plant. Then he'd return to Mexico at 11 p.m. and then some time in the early morning hours of the next day they'd load it again.

The trucks involved in this operation were of the flatbed bobtail [*tortón*] type that had about half the capacity of larger flatbed trailers—or approximately 6,000 brick per load. Cain estimated that about 12,000 crude brick per day were hauled to the Relámpago plant from several Mexican sites, including Las Flores, between Nuevo Progreso and Ciudad Rio Bravo. According to Cain the Relámpago plant's crude brick hauling was steady in 1954 and 1955 but ended in 1956 "because they kept picking up his laborers . . . they were all wetbacks."

In the February 1955 *Rio Grande Valley Yellow Pages* a new listing appeared under the heading of "Brick–Building", namely, Rio Importing Company with offices in Hidalgo and McAllen specializing in "*Mexico Hand Made*." The Hidalgo office was, in fact, located at a brickyard just 100 yards west of the Missouri Pacific Railroad right of way northwest of U.S. Highway 281 in an area known as Los Jardines or *El Jardin* [The Garden] because it was also the site of a nursery that sold plants [Map 2, black dot to left of 'Hidalgo' marker]. Handmade brick were made at this site in a kiln that had been constructed on a slope so that its top portion was a ground level and the firing chamber below at the bottom of the slope. This kiln was wood–firing prior to 1954 when a gas line was installed and two new kilns were constructed. There is ample testimonial evidence that members of the Soto family, Andrés and his sons Eduardo and Guadalupe, built and tended these kilns, and were the principal brickmakers at this plant on those occasions when brick were molded there. So, 'El Jardín' in Hidalgo became the site of Rio Importing Company's *maquila* operation in Mexican handmade brick. To my knowledge, Rio Importing was first in the lower Rio Grande region to specialize in the production and sale of Mexican handmade brick according to a *maquila* plan.[10]

Rio Importing Company's main investor was Trinidad Herrera, member of a prosperous Reynosa family of rural and urban landowners, who owned a large cattle ranch *La Llorona* outside of Reynosa. He did business with Reynosa's brickmakers by selling them firewood for their kilns, helped resolve problems related to immigration or customs at the Reynosa–Hidalgo bridge, and was well connected in cross–border business circles. His wife, Adelina, owned a store near the bridge referred to popularly as "*la tienda del rio*." Also, involved in the Rio Importing Company was Lester Tisdale of Pharr, an entrepreneurial ex–Border Patrol officer who became an employee of U.S. Customs, and in the early 1950s transitioned out of U.S. Customs into business with Herrera. At the time, Tisdale already had established his own separate business as a home builder in the Valley, and had purchased and used Mexican handmade brick in several of his projects. This is how he met a Houston–based brick dealer and importer of Mexican handmade brick named Bill Cypert, who became a partner with Tisdale in the brick business; and Magín Villaseñor, a Reynosa–based trucker who was hauling brick from Saltillo for a dealer named Armendáriz, and who Tisdale financed to start producing unfired brick in Reynosa to be fired in the Hidalgo plant.

Another person linked to the Hidalgo brick plant, perhaps to its pre–Rio Importing Company period, was a trucker from Mission, Texas named Ramiro Rodríguez who was Trinidad Herrera's brother–in–law. Prior to 1954, he had sold Mexican handmade brick to Cypert and, perhaps, also to Tisdale. When Cypert and Tisdale became partners, Ramiro became their hauler. After Tisdale's departure in 1957, Rodríguez and Cypert became partners in the Hidalgo plant.

Before 1954 Tisdale had been buying handmade brick, already fired in Saltillo, from a Texas–Mexican intermediary named Armendáriz who hired Villaseñor to transport brick from Saltillo to Hidalgo. Tisdale and Armendáriz had a falling out, and Tisdale [together with Cypert] decided to establish his own brickyard in Hidalgo with kilns to fire crude brick. So, we have a rather curious situation where Trinidad Herrera financed Tisdale, and Tisdale financed Villaseñor to produce crude brick in Reynosa to be fired at the newly established brickyard in Hidalgo that was apparently owned by Herrera. Villaseñor was predisposed to do this because on one of his trips to Saltillo for Armendáriz he had spent several days helping out some under–producing brickmakers so that he could come back to Hidalgo with a full load. During that experience according to Villaseñor, "the idea of someday having a brick plant was born inside me like a parasite" [*Se me nació el gusanito de algún día tener una ladrillera*]. Thanks to Tisdale, Cypert and Herrera, Villaseñor got his chance, and in a few years he would become one of Reynosa's largest and most respected brick producers.

Whereas by 1954 Cypert became, and would remain, a specialist in the brick business, Tisdale and Herrera had other business interests they pursued through Rio Importing Company—pineapples, coffee, and cement among other commodities. The export of coffee was prohibited by the Mexican government at the time [not by the U.S.] but Cypert remembered that bundles of coffee were hidden on trucks loaded with brick and smuggled across the bridge by the Rio Importing Company.

It was the cement business, however, that led to Tisdale's bankruptcy. To make a long story short, he managed to corner the output of Monterrey's largest cement companies for sale in the Texas market. This led to a shortage of cement for the Mexican construction industry from Monterrey to the border that elicited a protest by Mexican builders to the Department of Industry and Commerce in Mexico City and, ultimately, to a temporary embargo by Mexican Customs on building materials exports. The combined effect of the embargo and the cancellation of a large order from a West Texas client, left Tisdale with tons of bagged cement that were impounded and eventually turned into concrete at Roma and Rio Grande City ports of entry. Bankruptcy resulted and Tisdale moved to Louisiana where he got reestablished in the business of importing Mexican handmade brick. During the mid-1950s when Tisdale concentrated on the cement business, Cypert and Ramiro Rodríguez had, for all practical purposes, taken over Rio Importing's Hidalgo brick plant. They continued to run it as a *maquila* operation for a period after Tisdale's move to Louisiana, presumably still with the backing of Trinidad Herrera.

As often happens in cross–border trade, unexpected problems, bureaucratic or otherwise, arise to interfere with business as usual. In the case of the Hidalgo brickyard, the U.S. Department of Agriculture decided to halt Cypert's and Rodriguez's shipments of unfired brick from Reynosa due to an arbitrary ruling that crude brick were made out of raw dirt rather than processed clay, and raw dirt was regarded as a source of nematodes [parasitic worms] that threatened Valley agriculture. Cypert and Rodríguez were frustrated in their efforts to overcome this temporary embargo and, after a short period of experimentation with making handmade brick on site at their Hidalgo plant with Mexican day laborers that proved unrewarding, they became strictly importers of already fired handmade brick from several small plants in Reynosa.[11]

The story of the *maquila* strategy in the border handmade brick industry did not end with the demise of the Hidalgo plant. In the early 1960s, Howard Rogers and Clay Moffett were partners in a new brick plant located at the intersection of Dove Road and N. Tenth in McAllen. The plant featured a 131 foot long, gas–fired tunnel kiln with cars that would hold 1,000 brick each and turn out 1,000 brick per hour. Somewhat ill–advisedly, right in the aftermath of the forced repatriation of undocumented Mexican workers in the 1950s, Rogers and Moffett attempted to produce handmade brick at this plant to fire in their kiln. Not surprisingly, they were unable to do this successfully, and then operated for a time on a *maquila* basis, importing unfired brick from Reynosa plants.

This ill–fated new plant went bankrupt and in 1963 was acquired by Carl Scalese who partially repeated the experience of the original owners. In his words (interview 7/1/93): "I decided to go into Mexico and make the brick there and fire them here but I wasn't able to do it. They could never produce enough." More specifically, Scalese leased Raúl Valera's idle plant near the PEMEX [*Petroleos Mexicanos*] refinery for about two years with the idea of manufacturing crude brick to fire in his McAllen tunnel kiln but ended up attempting to also fire crude brick in Mexico. He found "the men that had been working for Valera were just staying there at the plant," and continued: "I don't know how they were existing. When I got there they were all dying of hunger. They had little shacks there." He was shocked by the fact that, "they were giving their babies beans instead of milk because they couldn't afford to buy the milk."

Scalese's post–*maquila* experience in Reynosa failed due to union trouble and to bad luck. In his words: "When the union started giving me all of that trouble, I decided to close it down. The union struck me. They closed off all the entrances, had their red banners up, and all that." And he further explained: "We had open, updraft kilns. I had one filled with about 300,000 brick. We got a big rain and it just completely destroyed all of them." This combination of factors forced him to shutdown and to quit the Mexican brick business in 1965.

As these case histories illustrate, handmade brick production in the lower Rio Grande corridor is inherently a two–sided affair, typically mating U.S. capital with Mexican labor. Even when green card commuter labor was employed over the years, it proved to be too unreliable and too expensive vis–à–vis labor in Mexico to allow plants on the Texas side of the border to operate competitively. Since most of the brick molding labor employed in Rio Grande Valley plants up to the 1950s was undocumented, the plants were subject to periodic raids by the Border Patrol resulting in severe disruptions of their operations. The *maquila* option was adopted by plant owners as a strategy to resolve the labor problem by transferring handmolding operations to satellite fields [*campos*] or plants in Mexico. In short, the political economic cards were stacked against handmade brickmaking on the Texas side of the border and favored Mexico–based industry.

Contrasting Approaches to Mexican Border Brickmaking in Global Capitalism

Tamar Diana Wilson, in her study of brickmaking in Mexicali, is critical of my flexible accumulation–inspired approach to understanding the border brick industry that emphasizes an endogenous dynamic of capitalist enterprise development (Cook 1998: 283). Instead, she argues that brickmakers on the U.S.–Mexico border who she categorizes, for the most part, either as petty commodity producers or pieceworkers, are "subsidizing capitalism." They do so directly, she contends, to the degree that their brick are "used in the construction of commercial and industrial parks and structures and are acquired at a cost lower than could be produced by a formal–sector firm operating according to profit–making principles," and indirectly through the use of their brick "in self–built or semi–self–built housing" that "lowers the pressure for a higher wage among the formal–sector workers who build and occupy them" (2005:162; cf. pp.50, 75, 107, 135).

Wilson does not empirically demonstrate these two kinds of subsidies in terms of actual commodity–value flows. Rather, they are simply posited as logical outcomes of the particular subaltern location of brickmakers in the class structure of capitalism–in–the–abstract or what she refers to as "the overarching capitalist system" (2005:13). Her entire argument rests upon a controversial dualistic view of that system as encompassing a *formal sector* [capitalist enterprises] and an *informal sector* [petty commodity–producing or other household enterprises], a view that leaves petty capitalist enterprises in limbo. Many researchers, including myself, have rejected dualistic thinking in favor of an approach to the capitalist economy that emphasizes its complexity by branch regarding scale of and relations between enterprises, its interior and exterior ties to other sociocultural fields, as well as an analytical focus on the relations of commodity value distribution in specific enterprises and branches (Cook and Binford 1990:Ch.1; Cook 1998:279–283; Cook 2004:Ch.8; cf. Kearney 1996).

Artisanal brickmakers, whether in Oaxaca, Baja California, Chihuahua, Coahuila, Tamaulipas or elsewhere in Mexico are producers, whether wage workers or worker–owners/renters, whose commodities are sold with the expectation of profit in competitive, interconnected markets in which varieties of building materials are available at different prices or are salvaged from waste. In these building materials markets prices are typically negotiable, even for products sold by the largest enterprises, and my experience has been, in Oaxaca and in Reynosa, that selling prices of handmade brick are by no means fixed from transaction to transaction but, rather, vary significantly in terms of the identity of the buyer, the quantity sold, buyer knowledge of the market, the necessity of the seller, and so on. For example, in Reynosa handmade brick of any size sell for a higher price per brick in small lots than they do in large lots; and do so both export and domestic markets.

In short, there are many relevant issues that can be explored about commodity value transfers and their supporting relations of production and distribution in the Mexican handmade brick industry, but subsidization ranks low on my list of them.[12]

Conclusions

The meandering Rio Grande provides resources necessary for brick production by tough, skilled artisans who excavate and prepare clay, mold it with wooden molds, then dry and fire the brick. Brickmakers were always of Mexican origin, but during the twentieth century handmade brick production ended on the Texas side of the border and became a Mexico–based industry, although for a time the industry operated on a *maquila* basis. Handmade brick retain a niche in the twenty–first century Texas masonry construction market due to their irregular and rustic appearance that cannot be simulated through machine production. Under conditions of global market capitalism, handmade brickmaking enterprises are best understood as competing in the building materials market with other commodity–producing enterprises rather than as subsidizing particular sectors of the wider system.

Despite the ups and down of the business cycle, the confusing shifts in exchange rates, and a concerted campaign by large–scale Texas brick producers to eliminate competition from Mexican handmade brick in the Texas market during the second half of the twentieth century (Cook 1998: Ch. 12), it is noteworthy that handmade brick production on the Mexican side of the border has survived into the twenty–first century. That it has done so is a tribute to the business savvy and perseverance of the surviving producers but, most importantly, reflects the persistence of a demand in the mainly upscale Texas custom building market for brick with a rough, rustic, natural adobe look.

Notes

1. In my interview with Ed Stout, president of Acme Brick Company in 1992, a job he had held for nineteen of his forty-three years with the company, he never mentioned the topic of the 'El Ford' line in his extensive comments on Mexican brick or on the round beehive kiln soft–mud facility they were operating at the time in north Texas. He did, however, discuss extensively his experience with and views about Mexican handmade brick. He had made several trips to the border and to the interior of Mexico, both for the purpose of possibly buying or partnering with Mexican plants but also simply to observe Mexican brick culture or, in his words, "Just to see and have a laying on of the hands and know more about my competitor." He emphasized that his company "looked at Mexican brick as we look at brick from anyplace else, they're competitors . . . we look at it as a competitive product." He also considered it to be inferior quality–wise to U.S. production brick due to the raw materials used in making it, the shortcomings of the firing process, especially when wood was the main fuel and kilns were of the simple scove type, and the lack of quality control. He admitted that over the years "we have made attempts to buy and resell Mexican handmade brick. It's difficult to do and we were never able to get consistent service." He further emphasized that "if something goes wrong with a product Acme Brick sells, our customers expect Acme Brick to take care of it. If you are dissatisfied, we are going to fix it. The majority of people that sell Mexican brick say, 'Well you bought 'em and they are yours.' That influences what we do." Contrary to the precedent set by O'Neil Ford, Stout's experience with architects had been that they would "never put their reputation on a building that had Mexican brick on it." For that reason, Mexican handmade was "pretty well restricted to the residential market." He admitted, however, that "there will always be Mexican handmade brick used in Texas" and that it could become a major player in the market if Mexicans could produce a "high quality, unique brick that would carry the freight." He did not think that the Mexican handmade he was familiar with made the grade from an aesthetic standpoint. But it could do so if it were unique and different—different colors, different sizes, and so on. Hence, Stout provides an explanation for the current success in the Texas market of *Ladrillera Reynoso* documented in chapter 9.

2. Acme ceased to be a Texas–owned enterprise in 2000 when it was acquired by Warren Buffett's Berkshire Hathaway conglomerate.

3. In common parlance and in most of the academic literature, 'manufacture' is a cover term for factory production, an unfortunate usage that does violence to the prefix 'manu–' that comes from the Latin word 'manus' for hand. In a strict sense, the term 'manufacture' should refer to labor–intensive production processes in which the instruments of production are tools of manual labor, not machines. I prefer to use the term 'machinofacture' with reference to machine–based production processes. Given the ambiguities associated with the use of 'manufacture', I will use the more general term 'production' with specific modifiers like 'handmade' or 'machine–made' (cf. Bottomore et.al., eds., 1991:331).

4. River alluvium is not the sole source of raw material for brickmaking in the lower Rio Grande corridor. In geological time wind was undoubtedly also a factor, together with river course uplifts and shifts, in creating loamy/clayey sediments of commercial importance for brickmaking referred to as loess that is present in the Laredo area and probably downriver as well. In Great Britain this material is called 'brickearth' and was important for brickmakers there as well as in such diverse areas as the Rhine Valley in Germany and the Yellow River Valley in China (see entry for 'Loess' in Wikipedia). It was, for example, a component of the raw material used in late nineteenth–century handmade brickmaking in Arkansas, described in a geological report of the period as follows (Call 1891:161): "The clays used for brickmaking at this yard [Cook's in St. Francis County] are all surface clays, the loess from the hills, mingled with some sand...and mixed with the humus or top soil." So long as this material is clean, low in content of mineral nodules or gravel, it comprises a prime raw material for brickmaking and was put to use by late nineteenth century brickmakers like A.H. Thaison in Laredo (Adams 2008:140; cf. Cook 2010). See Gurcke 1987:3–4 for a concise discussion of 'clay.'

5. Map 2 has been adapted from Maps 191/192 produced by the United States Geological Survey in cooperation with the Department of Treasury, U.S. Customs Service. It embodies photoimagery rectified by optical scanning from a color infared aerial photograph taken July 12, 1982 by the *Dirección General de Geografía del Territorio Nacional* (DGGTN), México D.F., Mexico. Location and lettering of brickmaking sites and *Colonias* was done by Ray Blanchette, a draftsman in the University of Connecticut Publications Department.

The entire municipality of Reynosa is administratively divided into '*colonias*' or named residential/neighborhood areas. The designation implies human settlement but population densities and settlement histories vary between them. Many *colonias* are fully urbanized and densely settled whereas others, particularly those with *ejidos*, are periurban, semi–rural, and less densely populated. The term *colonia* assumed a special meaning in the context of "regional irrigated development projects built by the government in the river valleys of northern Mexico after the Revolution" to refer to settler communities, comprised of repatriated migrant workers or other landless workers (Walsh 2008:7 et. passim). Some of Reynosa's *colonias* did apparently involve settlement sponsorship by the government linked to economic development. So-called "*colonias populares*" typically are squatter settlements that are grudgingly recognized but marginalized by government policy.

6. As mentioned in the Preface, documented by Walsh's (2008) pioneering study, the Tamaulipan side of the riverscape, the Lower Rio Bravo Valley [Valle Bajo Río Bravo], starting in the late 1930s became the focus of a large federally–sponsored irrigation, agrarian reform, repatriation, and colonization project to grow cotton through a combination of private and public sector enterprises. This coincided with a policy turn by the Cardenas administration from emphasis on the collective *ejido* toward a colonization/repatriation model favoring small property owners (Walsh 2008:136). By the 1950s the entire region had been substantially transformed—although with a mixed agrarian class structure of *ejidatarios*, *colonos*, and *pequeños propietarios*. Complications in the cotton market, significantly reflecting U.S. policies toward cotton subsidies and importation but also including pests, plant disease and soil depletion, forced a shift away from cotton toward corn and sorghum in the 1960s. In any case, the political economy and political culture of the entire region was impacted by this project, or by related federal

government land reform and development policies. So, the riverscape for the development of the greater Reynosa brick industry encompassed regional agricultural development as well. More recently, agriculture on both sides of the lower border has been severely impacted by drought and the lack of water for irrigation (Walsh 2004).

7. A concise characterization of the Rio Grande is as follows (HARC 2008:1): "The Rio Grande is the fifth largest river in North America. It runs North–South from its source in the Colorado Rocky Mountains to El Paso, Texas. From there it turns South–East and, for over 800 miles, serves as the border between Mexico and Texas. Due to intensive human development and water use in Colorado, New Mexico and West Texas, the Rio Grande runs dry south of El Paso during much of the year. As a result, the Rio Grande/Rio Bravo below Fort Quitman, Texas to the Gulf of Mexico has effectively become a separate watershed. Its climate is arid to semi–arid, and vegetation and wildlife consist of Tamaulipan scrub and chaparral desert ecosystems." The Lower Rio Grande/Rio Bravo Border Region, that comprises the final 150 miles of the river and the surrounding watershed below Falcon Reservoir in Zapata County below Laredo, is a dependent sub–region in hydrological terms of the Rio Grande Basin or watershed beginning below El Paso at Fort Quitman. However, releases from the Falcon Reservoir comprise the major component of the river's flow equaling about 95 percent of its water (ibid., p.3).

8. This statement was made during a tape–recorded roundtable discussion held in Edinburg, Texas on June 9, 1993 at the Echo Motel conference room by the Border Brick Industry Project. The statement is quoted verbatim from a transcription of the roundtable proceedings made by the author. Regrettably, Robert Runyon's photos of the Casa de Palmas hotel were taken just as it was nearing completion and immediately after completion, so he was not there to photograph the phase of on–site brickmaking.

9. Without burdening the reader with a bewildering variety of types, shapes and sizes of molded clay products, the basic distinction in the border brick industry is between the so–called *adobón* that typically measures two or three inches in thickness by five inches in width by ten inches in length [although the largest variety measures four inches in thickness, eight inches in width and sixteen inches in length]; and the *ladrillo de ocho o nueve* [common or standard modular brick] that measures two or three inches in thickness by three to four inches in width by eight or nine inches in length. *Adobones* are molded in one, two or four compartment molds, whereas *ladrillos* are almost always molded in six compartment molds.

Apparently the box molds used in border Mexican brick culture are less common worldwide than the open type without closed bottoms because "the suction created in the bottom of the mould makes it difficult to turn them out" (Campbell and Pryce (2003:15). It is to overcome this "suction" that Mexican *hechureros* always sprinkle sand into the bottom of the mold prior to dropping in the clay. Gurcke (1987:39) observes that box molds were in use in Jericho by 3000 B.C. A rectangular open frame mold is used in Oaxaca brickmaking where it is positioned on the drying ground by the molder and filled from a bucket (Cook 1984:68–76). By contrast, a square open frame mold is filled with a shovel in Oaxaca adobemaking (Cook 2004:223–225).

10. Anita Soto, *La Paloma*, who spent a great deal of time moving back and forth between family houses in Hidalgo and Reynosa during the 1950s, provided crucial details about her family's role in the Hidalgo plant, and also about Ramiro Rodriguez and the Herreras, Trinidad and Adelina, in three separate interviews conducted by the author in

1993, 1995 and 1996. She specifically stated that "my father was the one who constructed the ports [*boquillas*] in the kiln in 1950."

11. Soil, defined by the USDA for Customs purposes as the "loose surface material of the earth in which plants, trees, and shrubs grow," is still prohibited entry to the United States from Mexico unless accompanied by an import permit.

12. There is a host of problems that stand in the way of empirically demonstrating the validity of Wilson's subsidization thesis. The following immediately come to mind. First, by her own admission, some of the brickmaking enterprises she has studied regularly employ wage labor and are classified by her as petty capitalist (e.g., 2005: 12, 105, 134). She also finds that increasing numbers of these successful micro enterprises are beginning to pay taxes and social security for themselves and their workers, hence "formalizing" their capitalist status and driving up their production costs (2005:161). Who, then, is subsidizing whom? Second, fired brick are not especially cheap in relative terms and, in fact, are more costly than unfired adobe blocks or other scrounged materials used in self–building or semi–self building in arid Mexico's *colonias populares* or shantytowns. So, how precisely, can they be seen as contributing to cheapening the exchange value of labor power [i.e., the labor reproduction cost] for capitalist employers? Third, building materials like brick are only one ingredient in housing or other building construction, and housing is only one of many factors determining the reproduction cost of labor power—certainly less important than, for example, food. How can fired brick, then, be meaningfully factored into to an equation, or operationalized in terms of 'wage goods' calculations, to demonstrate "capitalist subsidization" or an impact on producing a cheaper wage goods bundle that lowers [or keeps low] labor reproduction costs so as to assure higher profits? In my opinion, this would be an econometric exercise in futility.

Chapter Two
Foundations and Dimensions of Border Brick Culture

Before 1848, with the exception of a few settlements like Matamoros, Reynosa, and Camargo located south of the river, there was scant brick masonry construction and, consequently, relatively little brickmaking along the lower border. North of the river, only Brownsville and Laredo had significant numbers of brick buildings by 1848 and most of these were built from brick made by local practitioners of Mexican brick culture. Before the river was bridged in the 1910s and 1920s, brick handmade in Mexican towns like Reynosa were transported by barge or mule– or ox–drawn carts across the river to Texas to meet a local demand that was not being fully supplied by producers on the Texas side.

After 1900 as the Rio Grande Valley urbanized, Mexican brickmakers from Reynosa and other Mexican border towns provided the manpower for brickmaking to expand on the Texas side of the river, either in temporary plants established at or near particular building sites, or in more permanent commercial plants servicing multiple building projects locally and regionally. With few exceptions, in both types of brick plants in Texas, the entire process of making brick exclusively involved Mexican–descent personnel: the anonymous molders and movers of brick into and out of the kilns, and the plant foremen [*mayordomos*] who supervised brick molding and drying, construction and maintenance of kilns, and the all–important kiln firing and cooling–down process.

Overview of Regional Brickmaking History

Whereas the pre–1900 history of brick production and construction in the Rio Grande Valley revolved around a combination of military, religious, ranch, and urban construction axes, the post–1900 history is dominated by the demographics and economics of urbanization linked to capitalist development of agriculture, infrastructure, and central places. Before 1900, brick production, with the exception of a handful of urban areas like Reynosa and Matamoros was undertaken for specific building projects, often on–site, rather than for sale. After 1900, the development of major town sites in the Rio Grande Valley like San Benito, Harlingen, Edinburg, Mercedes, Weslaco, McAllen, and Mission involved considerable brick construction and resulted in the establishment of commercial brick plants. Most of these new plants were active for a decade or less, some were active for two decades, and a handful appear to have survived for as long as three decades (see Cook 1998: chapters 2 and 4).

Manuel de la Viña, who was mayor [*alcalde*] of Reynosa in 1850 (Herrera Pérez 1989:75), is credited with having "sponsored the construction of a major *ladrillera* during the 1840s" (Boyd et al. 1994:73). The 1837 Reynosa census shows that up to that time brick was not a significant building material there since buildings were constructed either of stone, wood or *carrizo* [thatch] and *zacate* [corn stalks] (Herrera Pérez 1989:127). The de la Viña–sponsored *ladrillera* may have been the earliest regular source of brick for Reynosa masonry construction, although historical records in the Reynosa archive also refer to a *ladrillera* of Epifanio Domínguez in 1850 (Minerva Morales de García, unpublished manuscript).

Census data indicate that Reynosa's population remained essentially stable from 1837 with 5,346 inhabitants to 1860 with 5,020 inhabitants, showing marked growth only from 1860 to 1884 when it had a population of 6,692 (Herrera Pérez 1989:131). Consequently, the 1850 demand for *ladrillos* in Reynosa must have derived from the improvement of commercial and public buildings and private houses along with the construction of new ones.

The archival record discloses that for several decades of the nineteenth century Manuel and other members of the de la Viña family were associated with a ranch located in a zone of historic meanderings of the Rio Grande that today includes land immediately to the east of the recently constructed Pharr–Reynosa International Bridge. Within the landholdings historically identified with this ranch, known as *Rancho El Capote*, archaeological excavations yielded evidence of brickmaking and brick construction (Boyd et al. 1994) [Map 2, black dot on upper right]. The most promising connection in the El Capote Ranch area to the wider history of brickmaking in the Lower Border focuses on the Guajardo–Vela site which had a small *ladrillera* that, according to the authors of the site report, "could have been built and used in the latter half of the nineteenth century" by someone other than its then owner Pedro Guajardo who, they note, "could not have used the kiln himself until the 1890s" (Boyd, et al. 1994:114; Boyd 1995). It is further documented that Guajardo did, at some point, build a brick factory on this site that operated until around 1922 (Boyd et al. 1994:101; Boyd 1995); and that the site was purchased by Eduardo G. Vela in the early 1940s. Between 1946 and 1948 Vela "hired Mexican *braceros* to make brick under the supervision of the Soto and Gallegos families" (Boyd, et al. 1994:115).

Another Hidalgo County *ladrillera* where various members of the Soto family worked, and that developed commercially beyond the minimum level achieved at San José Ranch in Cameron County, was at Toluca Ranch near Progreso [identified as TRBY in Map 3]. Toluca Ranch was founded in 1880 by Florencio Sáenz but its brick plant probably did not get started until the turn of the century when several buildings, including the main house, a store, and a small gothic chapel, St. Joseph's, were built of handmade brick produced on the ranch (Isbell, ed., 1994:125). The ranch *ladrillera* had two mesquite–burning kilns located at different sites, one near the main house and the other close to the river bank, and a largely resident ranchhand/brickmaker labor force some of whom "had their own *parcelas* that they worked themselves" through a *por partes* [rent in kind] arrangement with Saenz (author's interview with S.G. Fernández 1993). This plant clearly evolved beyond production for own–use since several commercial buildings and homes in Mercedes, Weslaco, and Progreso, among others, were built from Toluca Ranch brick. S.G. [Jimmy] Fernández, Florencio Saenz's grandson, recalled "seeing them take loads of brick on a wagon from here . . . toward Progreso Lakes" where "homes . . . were built with brick from here." The Toluca Ranch *ladrillera* had ceased to operate by 1940.

Although poorly documented, it seems clear that after 1900 brick production in Mexico also experienced a substantial boost from the construction boom in the Rio Grande Valley. Morales de García (unpublished manuscript) writes: "When the Anglo–Saxons who came around 1900 . . . needed to build houses, brick were taken by barge . . . The houses in Hidalgo . . . 1870s style . . . were made of Reynosa brick . . . for a long time, until the international bridge came in 1926." Nevertheless, in the Mission area of the Valley large–scale construction projects, like the chimney for the water pumping plant and the novitiate for the Oblate order of Catholic priests, stimulated local brickmaking that led in the case of Guenther Weiske to the establishment of a commercial plant [see below].

By the turn of the century there were reportedly as many as fifteen *ladrilleras* in Reynosa, including the plants of Melardo González. González had a ranch in Reynosa in what is today *Colonia del Prado* that encompasses the *Zona Rosa* a commercial tourist district, where during the first two decades of this century he also had *ladrilleras*. Again, according to Morales de Garcia: "This gentleman was very entrepreneurial. And one finds it there in the Municipal Historical Archives, documents say that don Medardo had sources of employment for people who made brick, that he had sufficient brick plants to meet the needs of the times, the first decade of the twentieth century." Her narrative continues:

> On the banks of the river there was an abundance of sandy clays. . .an abundance of raw material to make brick. For that reason the old brickmakers located along the banks of the river. There were a lot of them . . . an example, don Medardo González built several kilns along the river and the clay banks, they gave employment to our people. They sold in the Rio Grande Valley.[1]

Another *ladrillera* that was in operation in Reynosa during the 1920s was under the proprietorship of Antonio McDonald Castillo who married Carolina Longoria of the *Hacienda Longoria*. The *ladillera* was located on hacienda property in the riverine zone and, hence, was a forerunner of the many *ladrilleras* established in that sector in subsequent decades. As a result of the agrarian reform in the late 1930s and early 1940s, the *hacienda* was divided into private plots [*pe-*

queñas propiedades] and *ejido* plots [*parcelas ejidales*]; the area is identified today as *Ejido Los Longoria.*[2]

Also included in Morales de Garcia's narrative is an oral history anecdote from the early decades of the twentieth century that has several Americans running out of an Hidalgo building, perhaps the pumping plant, in response to a loud noise associated with the collapse of several brick kilns across the river caused by flooding. This was apparently a rather common occurrence in the days before flood control since the brick kilns [*hornos, obradores or cocedores*] were customarily located in close proximity to clay pits [*pozos, barrancos*] in the riverine zone. A prominent theme in the oral tradition surrounding brick production in Reynosa is that floodwaters ruined the kilns. Periodic river flooding plagued *ladrilleras* of that period like those of Medardo González, Antonio McDonald Castillo, and others remembered by Ramón Longoria from his boyhood days in *Rancho Longoria.*

Longoria grew up on his father's ranch. In my 1993 interview with him, he remembered that "when I was a boy, there were brick plants around my father's ranch and they made mostly eight–inch long brick and they fired them in kilns using firewood and they came out a yellowish color. Brick that were not yellowish in color were discarded. Those yellowish brick last more than 100 years in a building and they are well fired."

In 1935, around the time that Longoria saw brickmaking in Los Longoria, Manuel Lozano, a migrant to Reynosa from rural Nuevo León, started working at the age of fifteen in a La Copa plant owned by Alberto Gómez. He described the plant as just beginning but as having six or seven brickmakers, four setters/loaders, two truck drivers and two trucks, as well as two kilns, with capacities between twenty-five thousand and thirty thousand each, that alternated between loading and firing. Lozano, after passing through a period of learning the work, was able to make an average of 1,000 eight–inch brick in a half–day for which he was paid around six pesos [less than two dollars]. This compared favorably with the same wage that was then being paid for local agricultural labor for a workday that lasted from sunrise to sunset He also produced foot–long brick called *barras* for which he was paid more. The brick produced at the plant were marked *MEXICO,* which means that they were made for export.

The Soto Family on Both Sides of the River in the Nineteenth and Twentieth Centuries

In casual conversations about the history of brickmaking on both sides of the lower Rio Grande border the Soto surname inevitably is mentioned as representative of the industry and the craft—the Sotos are widely associated with specific brick plants as well as with brickmaking itself. The Soto family has participated in itinerancy and cross–border circulation for more than a century. Beginning around 1880 the patriarch, Francisco Soto Salazar, born around 1855 in the Hacienda San Salvador near Linares, Mexico, who learned brickmaking in his youth, took up residence area at the Andrés Chávez ranch near Hidalgo, Texas. Soto's six sons, all of whom became brickmakers, were born at the Andres Chávez ranch. His youngest son, Andrés, born in 1892, and Andrés' sons, Guadalupe and Eduardo, practiced the brickmaking tradition of their patriarch up and down both sides of the river for many decades of the twentieth century, and have assured the Soto surname a prominent place in lower Rio Grande/Rio Bravo industrial history.

Aside from his possible involvement in brickmaking at the Guajardo site during the 1880s and 1890s, Francisco Soto also worked at *ladrilleras* in Starr County, Los Ebanos, and El Granjeno until ca. 1915. During the 'San Diego revolt' when the "armed insurrection of Texas Mexicans and its brutal suppression by Texas Rangers . . . turned the Valley into a virtual war zone during 1915–1917" (Montejano 1987:117; cf. Zamora 1993:81–84), Francisco Soto fled with his family to Reynosa where he and his sons established several *ladrilleras* in the riverine zone [*zona ribereña*], apparently with government support. In the words of Anita Soto, "my grandfather took possession of land along the river, the government helped them establish brick plants." Andrés returned to Reynosa with his father in 1915 and worked with his father and brothers for a time.

Alfredo Flores Sotelo, who was ninety years old when I interviewed him in 1995, remembered working with Francisco, Andrés, Atilano, and other Sotos in 1919 in Los Ebanos when he was fourteen. He emphasized that whereas he was an apprentice at the time, "They were already brickmakers when they came here. They already knew the work. They did not compare with the rest of us who began there because they were already brickmakers by trade." Regarding Francisco Soto, who in 1919 was elderly, Flores said: "it was his occupation . . . he set brick in the kiln prior to firing." Together with stoking, the setting of unfired brick in a kiln is a task reserved for the most experienced master brickmaker.

In 1924 Andrés Soto remarried back in Hidalgo, Texas several months after his first wife died. He lived and worked in a Granjeno plant, together with his new wife, Juanita Martínez, and his sons [from his previous marriage] Eduardo and Guadalupe in 1925 and 1926. José C. Anzaldúas, born in 1913 in Granjeno, remembered Andrés and Eduardo Soto, together with several other *ladrilleros*, living and making brick on a site immediately to the west of his father's property. The small frame house where the Sotos lived is still standing in Granjeno and several initialed brick made there have been collected by Mr. Anzaldúas. There were two kilns and a large clay pit, as well as an *estero* [settling basin] immediately to the south of Granjeno village; the clay pit was on the north shore of the lake, just beyond the brick plant [see Map 2, third black dot from left in upper left section].

Oral testimony and physical evidence from the Granjeno sites confirms the existence of three separate *ladrilleras* there within the Nicolás Bocanegra share of porción fifty-eight, controlled by the heirs of Porfiria García de Garza and by Irene G. de la Viña (*Hidalgo County Deed Records* Vol. 738:144 and Vol. 1114:362). Jacobo Garza is mentioned as a partner in at least one of these *ladrilleras* where members of the Soto family were employed. José Concepción Anzaldúas remembered that in 1918–1920, when he was six or seven years old, seeing abandoned kilns overgrown with brush on a site in Granjeno just to the east of his father's property on land that he remembers belonging to two sisters known as *las Garzitas*—presumably daughters of Jacobo Garza.

During the 1930s, 1940s and 1950s, Andrés and his sons also worked periodically in *ladrilleras* in Texas: Rio Grande City, Los Ebanos, El Granjeno, Hidalgo and, lastly, in Relámpago. This pattern of cross–border itinerancy can be viewed as a necessary consequence of full–time occupational specialization in an industry dependent upon the ups and downs of the construction market in a unique bi–national economic setting. As occupational specialists with roots on both sides of the border, the Sotos were able to avoid putting all of their eggs in one basket, so to speak, and ply their trade and seek their livelihood in a regional and binational economy rather than a more restrictive locally bounded one. Anita Soto's narrative of her father's and brothers' careers presents an evocative image of an itinerant, gypsy–like, patriband of *maestros–ladrilleros* for hire, with family over in Reynosa, moving from one Hidalgo County *ladrillera* to another as production waxed and waned. In her words: "Those poor guys roamed around like Gypsies, from one side of the river to the other, wherever they could find work." Yet the bottom line of this narrative is probably the interplay of choice and necessity implied in her statement that her father "went away from Reynosa to make his fortune, he had to go to the other side to work."

In 1926–27, Andrés Soto returned to Reynosa to start his own *ladrilleras* in *La Copa*, near what is today the bus terminal, and *Corrales* [see Map 2]. According to Raúl Valera, the Sotos had the best–known *ladrillera* in Reynosa at that time: "There was already another *ladrillera* [in addition to his father's] in that period owned by the Soto family. It was a large *ladrillera*. The Soto family is very well–known in this region. The city was small then." The Soto *ladrilleras* in *La Copa* were destroyed by the 1933 hurricane. Anita recounts the impact of the hurricane: "There was a big flood that finished all the brick plants. All of the patios were covered with water because the old settling basin was there." In the aftermath of the hurricane, Andrés moved with his family to another property on 1324 Guadalupe Victoria norte in Reynosa. Immediately upon arrival, according to Anita "Very large patios were made, [and] deep clay pits. He made his *ladrillera* there."

Around 1935, after the Soto family had relocated to 1324 Guadalupe Victoria Norte, a flatbed truck from the *ladrillera* of Raul Valera's father, Daniel, loaded with brick for Hidalgo turned over on a provisional pontoon bridge and fell into the river. This accident, recounted to me separately by Anita Soto and by Raúl Valera, earned Andrés Soto the nickname *submarino* [*submarine*] since he went underwater to help to put cables on the truck to pull it out of the river on the Hidalgo side. Unfortunately for Daniel Valera, the truck was not salvageable since it was covered with sand and the motor and differential were ruined.

Anita Soto was born in 1936 and was raised in the house and *ladrilleras* on Guadalupe Victoria Norte where in the 1990s her aged mother still resided. Her husband, Eugenio García Alvarado, hauled firewood to her father's *ladrillera*, and saw Anita from the time she was eight years of age loading brick on a cart. They were married when Anita was twelve years old. The Soto family's *ladrilleras* were in operation at the Guadalupe Victoria, off and on, for twenty years.

Soto located and described this neighborhood of family brick plants:

> From the elevated bridge, one has to walk some four or five blocks toward Aldama until you come to the Leonardo Garza de la Cruz school. Then you go down from the school to Escobedo street, all of that area is a large clay pit, very deep, where all of the *ladrilleras* were located. That's where the old *ladrilleras* of my family were located. On the corner of Escobedo and Gonzalez Ortega streets there was a large agricultural field. Just after the field was the house of my uncle Policarpio, he had a very large *ladrillera* on Escobedo street That's where uncle Carpio was injured after he fell into a well, drunk. To the north of Escobedo street toward the river bank, behind our property on Guadalupe Victoria street, were the *ladrilleras* of my father and my uncles. That's where I grew up in the mud [*soquete*]. In front of our house on Guadalupe Victoria, there was a large clay pit where my uncle Santiago, husband of my aunt Eustolia, had a *ladrillera*. And the son of aunt Cecilia, Francisco Gallegos, had another *ladrillera*. And my uncle Pancho Soto also had a *ladrillera* there. They filled in all of that property where the clay pits were in order to sell it. Now there are only houses on the property.

The itinerant lifestyle of the male members of the Soto patriline translated into an *absent father* in Anita's household of orientation. She remembers, as a young girl living in Reynosa in the house on Guadalupe Victoria Norte, making trips with her mother to a small store in Hidalgo, where her father had credit, to pick up groceries. Her father, she related, "did not come to Reynosa every week, he came every two or three months. He stayed in Texas to work, and we were the ones that crossed to pick up groceries where my father came and paid the bill."

In retrospect, the first stage in the intergenerational transmission of brick culture in the Soto family hinged on the relationship between Francisco and his six sons. Anita puts this in perspective: "All of them knew how to mold brick. They knew clays, which clays were best for brick. They knew how to set brick in the kiln. They knew how to fire. They learned everything from their father. My grandfather never worked in agriculture or anything else. He worked exclusively as a brickmaker." It is clear from this statement that Francisco Soto did not make brick simply as a complement to farm or ranch work or odd jobs. He was a full–time specialist in his craft; and he raised his sons to be the same.

It is important to note that brickmaking in the Soto family did not exclude females—at least in the relations between the second and third generations. Don Francisco's granddaughters, and particularly Anita and her younger sister Andrea, also learned brickmaking—in their cases from their father, Andrés—and worked in the family brickyards. Anita remembers that in her father's *ladrillera* on Guadalupe Victoria street, she worked as a young girl, together with her brother Guadalupe and her cousins Santos Castañeda, Natividad, and Daniél. In her words:

> As kids we started working in the *ladrillera* on Guadalupe Victoria street There my father taught me how to mold brick, how to haul brick on a cart, how to set brick in a kiln. He taught me how to load trucks directly from the kiln when the brick were still hot, 24 hours after firing started. To uncover them and load them, with bare hands, without gloves. My brother Lupe, my uncle Santos, and my cousins Daniel and Natividad worked with us. We had two small kilns, and made only size eight brick [eight-inches long, four-inches wide, and two- inches thick] to sell here in Reynosa.

Later, in the last of her father's plants in *Los Longoria*, again where brick was produced mostly for local sale, Anita helped to cart brick to the kiln and to set them for firing in order to reduce costs. During the final years of operation of her father's last plant, however, Anita was living with her husband and was involved in other business activities. She explains how her father became involved with the *ladrillera* in *Los Longoria*: "When the clay supply gave out on Guadalupe Victoria, my father came to *Los Longoria* on one or two hectares of land owned by Reynaldo Longoria. It was loaned to him, he didn't pay rent." Reynaldo's father, Alfonso, loaned her father money to run the brick plant and maintain his workers. The plant operated for seven or eight years until her father's death in 1964. Chapter 7 will discuss Anita Soto's career after 1964.

Guenther Weiske and Mexican Brick Culture in Madero, Texas

Only one enterprise in the Rio Grande Valley, namely, that associated with Guenther Weiske in the Mission area endured beyond three decades. Weiske, a farmer and landowner in Madero, got involved with handmade brick production by 1911 at a plant located near the river in *Rancho El Gavilán* on property belonging to his wife's family [Waaser], and historically associated with the Anzaldúas family who were the pioneer brickmakers there. Weiske's involvement was motivated by his interest in supplying brick for the large Oblate novitiate that was being built at that time on a site near the *El Gavilán* plant [see Map 2, upper left hand corner].

In 1923 Weiske and partner Hill Cocke decided to build a new plant closer to Mission and the railroad line on land that Weiske owned near Madero. Handmade brick was produced at the new plant but the emphasis was on producing machine–made brick with a J.C. Steele stiff–mud extruder acquired from Jack Decker, a local ice plant and machine shop owner who had intended to start a brick plant in Madero but decided instead to move his businesses to Mission. Weiske and Cocke's new Madero plant was called *Mission Brick & Tile Company.*

In 1930 Weiske and Cocke, as a strategy to shore–up their brick business during the Depression, arranged for a merger with the *Valley Clay Products Company* plant in Brownsville, as partners with J.C. Bennett and H.R. Stokley. At the time the *Valley Clay Products* plant was the Valley's largest both in terms of volume of output and variety of products that included roofing and drain tiles. After the merger, the Madero plant became secondary in the newly incorporated *Valley Brick and Tile Company* [VBTC]. The Brownsville plant had downdraft, beehive kilns that were gas–fired, whereas the Madero plant originally had updraft, wood–firing kilns. In the mid–1930s or so the Madero plant also converted to gas firing and also downdraft beehive kilns were constructed. Photo 11 [Appendix 1], taken by Robert Runyon in 1928, provides a panoramic view frontal view of the Brownsville plant adjacent to the railroad tracks with the beehive kilns and stacks of wood to fuel them prominently displayed. Other photos of that plant taken at the same time show the extensive clay pit directly behind the plant.[3]

Ironically, after the Weiske and Cocke families relocated from Mission to Brownsville and Harlingen, respectively, to be closer to the main plant, the infamous 1933 hurricane essentially destroyed it, and left the future of the VBTC dependent entirely on the Madero plant. In 1940 Hill Cocke acquired the *Crown Brick & Tile Company* located on 160 acres in Rio Grande City that also became part of VBTC. In 1943, however, Cocke sold out his interest in VBTC to Weiske who proceeded to further mechanize the Madero plant with additional extruders and a dry–press machine but also continued his involvement with handmade brick much of which was imported, both unfired and fired, from Reynosa area plants (see Cook 1998:104–106).[4]

Weiske's Madero plant produced brick continuously from 1924 until its closure in 1972. Its forty-eight year record of continuous production has only recently been eclipsed in the Lower Rio Grande region by one other plant that began mechanized production about forty years later than VBTC under the direction of Ramón Longoria, namely, *Ladrillerías Camargo, S.A.* in Ciudad Camargo, Tamaulipas. This plant was purchased in the 1980s by the García Garza family and operated until 2007.

Guenther Weiske's long career in brickmaking in Madero involved a continuing relationship with Mexican brick culture and its practitioners. As explained above, this began at his first plant at *Rancho El Gavilán* where he was wholly dependent on a Mexican–descent workforce including members of the local Anzaldúas family and others who crossed the Rio Grande daily to work for him. There is no question that Weiske learned the craft of brickmaking and kiln firing in that setting since his education and experience in Germany, prior to his immigration to the United States from Germany in 1910, had been focused strictly on agriculture. Weiske was enthusiastic about mechanization and technology; he was one of the first agriculturists in the Mission area to buy and use a tractor on his farm, and was a regional pioneer in producing brick with a stiff–mud extruder in his Madero plant. His modernist outlook, stimulated by Operation Wetback–sponsored Border Patrol raids at his plant, led him during the 1950s to an abortive attempt to produce simulated handmade brick with a soft–mud extruder purchased at the closure sale at Anderson's Relámpago plant. But Weiske found that it was cheaper to import Mexican handmade brick than it was to try to produce a simulated version of it with a soft–mud extruder. From 1923 into the 1960s he did business across the border, exporting some of his brick and importing Mexican handmade brick to supplement his plant's output.[5]

Weiske's Madero plant also had a sort of demonstration effect among several of Reynosa's entrepreneurial brick producers who admired its industrial technology. This is illustrated by Efraín Salinas' visit to Madero in the early 1950s to seek assistance regarding the installation and use of gas–firing kilns. He also imported and sold machine–made brick in Reynosa from Weiske's Madero plant. Ino Weiske, who I interviewed several times in the 1990s, remembered that one day at the plant his father introduced him to Salinas who at the time was thinking of converting his kilns to gas–firing. Salinas wanted to see the Madero plant's kilns in operation. Ino showed Salinas around the plant and then had him talk to his stokers. Afterwards, Salinas asked if he could send one of his employees over to Madero to observe the kiln firing process, or if Weiske could send one of his employees over to Reynosa to help in the installation of Salinas' gas–firing system. The Weiskes agreed to send over their relief stoker and setter, Fernando Flores, who was also the son of their plant foreman, to check out Salinas' kilns and to teach his employees to how to fire brick with natural gas.

The relationship between the Weiskes and Salinas lasted until the Madero plant was liquidated in 1971. The Weiske family had tried to sell the plant as a going business but without success, and were finally forced to auction it off piecemeal. It turns out that Salinas was the winning bidder for a stiff–mud extruder. Ino Weiske claimed that Salinas raised the price of his brick three or four dollars per thousand after the purchase because he never successfully installed the extruder in his Reynosa plant. One of the ironies of this story is that Guenther Weiske, on one of his visits to Salinas' plant in the 1950s, pointedly emphasized to him that he had the best machine for making brick, namely, "one that consumes beans" (Cook 1998:108–111). This was Weiske's way of expressing to Salinas his own frustration over not being able to get Mexican workers to make handmade brick in Madero due to the above-mentioned periodic Border Patrol raids at his plant, and to his lack of success in efficiently operating a soft–mud machine that was capable of producing brick with a handmade look.

Salinas, however, was having his own labor difficulties at the time and, like several of his colleagues in Mexico, aspired to manage a mechanized plant. This aspiration was confirmed by his 1971 acquisition of the Madero plant's stiff mud extruder. His lack of success in installing this machine at his Rancho Grande plant to produce machine–made brick for the domestic market probably reflected, as much as any other factor, his heavy involvement at the time in meeting a still burgeoning demand for handmade Mexican brick in the export market. It could also have reflected, as Ino Weiske noted, Salinas' way of keeping the machine out of the hands of a competitor.

The Weiskes in Madero and, later, the Nordmeyers in Rio Grande City, took mechanization further than any of their competitors in the lower border brick industry. Ironically, Harry P. Anderson attempted to do the same at his Olmito and Relámpago plants in the late 1940s and early 1950s—by introducing a soft–mud batching machine combined with a vibrating table that would essentially replace handmolders, the same machine that Weiske later bought from Anderson. Anderson's son, George, described his father's experiment:

> At Olmito, my dad got together with some of his cohorts and designed a tub–type batcher. He experimented and experimented and experimented, time after time, trying to get the right consistency of material. It was either too stiff or too wet Those guys that mixed it by hand had the magic touch. He tried a vibrating table so the molders didn't have to sit there and pat the clay in the mold and stick it off—the stick across the top of it gave one side the rough edge. I watched it. One day he said, "I give up." He put a lot of time and money into that batching machine and into that vibrating table. He tried two or three different things to make it easier to cut out some of the labor. He swore that there was a demon out there somewhere to make sure that Mexican handmade brick cannot be made mechanically. He wanted to perfect Mexican handmade brick but concluded that 'This stuff is meant to be made by hand—that's why they call it handmade brick'.

Over the years, however, mechanization was selectively introduced by Mexican producers. Indeed, as will be described in chapters 8 and 9, the two largest existent Reynosa plants, *Ladrillera Reynosa* and the new *Ladrillera Reynoso*, have mechanized clay preparation.

Music and Conviviality Among Itinerant and Repatriated *Ladrilleros*

Interviews with Anita Soto and others about the Soto family, and with Sebastian Reynoso, Lauro Rocha, Luciano González and others regarding their recollections of life and work in the Anderson brick plant in Relámpago, Texas, included several references to the importance of musical performance in the lives of *ladrilleros*. *Ladrilleros* either performed music for the enjoyment of co–workers or hired local musicians to entertain them on special occasions.

In the context of border culture, the connection between brickmaking and music is less paradoxical than it seems. The Texas–Mexico border region has a rich history of folk music, especially associated with rural and urban working classes [*clases populares*], that has evolved into genres known today as *ranchera*, *corrido*, and *conjunto*. Traditionally, this music was sung, accompanied or unaccompanied by instruments like guitar or accordion, and/or was primarily for dancing.

There is a historical record of this kind of folk music in the area south of Mercedes, the ranch community of Relámpago, which in the 1940s and 1950s had a brick plant. According to one source (Isbell 1994:86), in Relámpago: "Dances were held on a bare earth floor, slicked down with water. Five poles, one at each corner and one in the center of the dance floor, held lanterns. Benches were set all around. Musicians played the drum, accordion, and guitar, and sometimes the violin. Families brought their daughters to the dance . . . The sound of the *tambora* [drum] could be heard at the next ranch." This description is echoed in a statement by my informant, Fernando Flores, in his reminiscences about life in El Gavilán brickyard south of Mission around 1920: "All the brickmakers in El Gavilán on Saturdays danced to accordion by the light of lanterns hung on poles . . . There were many dances. People were happy." [6]

A great deal of testimonial evidence links the Soto family to music. According to Anita Soto, her father, Andrés:

> . . . liked to enjoy himself. After every kiln firing he invited everyone to go out to eat. He liked to eat fish, alligator gar was his favorite. Then he got drunk along with everyone else. My father's favorite song was *La Pajarera*. That was the song that they always sang. My father also liked the song *Te Vas Angel Mio*. He sang it when he brought musicians to the house. When my father came home from working in the United States, he would travel through the streets of Reynosa in a cart with musicians with two gallons of Barretiaga wine and they would sing *Te Vas Angel Mio* and *La Pajarera*. He hired itinerant musicians near the bus station and walked with them around the central plaza. They also came to the brick plant. My father was very fond of *mariachi* music, he liked it a lot.

The song *La pajarera* refers to a person who captures and sells wild birds as a street peddler but also has a more general meaning of a happy go lucky party person, jokester, and itinerant vagabond. With reference to her father and brothers, all of whom were *ladrilleros*, especially after the family brickyard in La Copa was wiped out by a hurricane in 1933, Anita stated: "The poor guys wandered around like gypsies, wherever they could get hired. And there were only a few brick plants." She said that her father told her that "The 'bird catcher' moved around just like he did, working in one place and another, from one side to the other—from Los Ebanos to Madero, from Granjeno to Relámpago, to Rancho Toluca, and Santa María; from Reynosa to *Estación Rosita* in Matamoros. Wherever my father went, his two sons Jose Guadalupe and Eduardo went with him."

During the 1920s Alfredo Flores, worked side by side with the Soto brothers, Andrés, Policarpio and Atilano, in a Los Ebanos brickyard. Aside from identifying the Sotos as *Reynosenses* and as being excellent workers, Flores characterized them as being very musical. As he expressed it: "They loved to sing and they sang well. They sang *a capella* without a guitar. They sang as a trio. I would go and sing with them. They sang a lot and they sang well."

In at least one Reynosa *ladrillera* in the 1940s, *ladrilleros* reportedly sang while they were making brick. Doña Minerva Morales remembers from personal experience that "when they were making brick, the brickmakers sang their song. It was in the 1940s that they sang. Their song had the following lyrics: 'A cinnamon–skinned girl with percale petticoats. My mother will be the sand, my father will be the mortar, my brothers the brick, soon the ocean will be bricked over.' [7]

Several *ladrilleros* reported that Cornelio Reyna in his youth, before he became a famous *ranchera* musician, worked as a brickmaker in Reynosa and, also, in Relámpago, Texas. Sebastián Reynoso, for example, who worked in the Relámpago brick plant in the late 1940s and early 1950s—before starting up his own *ladrilleras* in Reynosa—claimed that the musical group known as '*Los Relámpagos del Norte*' started there. That Cornelio Reyna, and Ramón Ayala, were brickmakers there, brickmakers and musicians who to entertain themselves and others in their spare time would pick up a guitar and sing. Reynoso's memory regarding Reyna has been confirmed by others but it may have been faulty regarding Ayala's presence at Relámpago.

Luciano González met Reynoso when they both were working at the Relámpago, Texas brick plant in the late 1940s and early 1950s. Given the fact that I interviewed Luciano after my interview with Reynoso, I asked him if he encountered either Ramón Ayala or Cornelio Reyna working there. He responded that he could not place them in Relámpago, noting that there were a lot of workers there, but did remember them in *Ladrillera Salinas* in Reynosa where he also worked. He specifically remembered that they used to walk from Salinas' *ladrillera* to play at a radio station located on the highway to Ciudad Rio

Bravo. The musicians González remembered from Relámpago were members of the Vela family, and particularly Rubén, who lived in a house along the military highway, not far from the brick plant. José 'Pepe' Vega, interviewed in 2006 in Reynosa where he was still working as a barber at the age of 84, worked in the Relámpago plant for several months in 1950, and confirmed that Rubén Vela was among those who sang in front of the *ladrillera* store.

Américo López López, sub–director of the Department of Commerce of the Office of Reynosa's Municipal President, in a December, 2005 interview confirmed the association between Cornelio Reyna, Ramón Ayala and brickmaking reconstructed above from interviews with *ladrilleros*. López, born in 1942 on a small ranch near Matamoros, came to Reynosa with his parents in 1949, and was raised in Colonia Longoria, then on the outskirts of the city with cattle ranches and dairy farms not far from where he lived.[8]

As a young man, López worked in pharmacies [*farmacias/boticas*] and had gained sufficient experience to open up his own business, with the support of his father, in 1960. His first pharmacy was located on the corner of Saltillo and Rio Mante streets in *Colonia Longoria* [shown as Los Longoria y El Olmo on Map 4] across from the Longoria School. Since this was one of the city's newest subdivisions, according to López, mostly lower–middle–class people lived there and were customers in his pharmacy. Many of his customers worked in *Ladrillera Tueme*. He remembers that Tueme provided his workers with space on plant property to construct shanties [*casuchitas*] to house their families.

Tueme's brick plant was on Praxédes Balboa Street, that paralleled the natural gas line [*linea de gas* or *gasoducto*], just a few blocks from Lopez's pharmacy. According to López: "All of those people came to me when someone in their family got sick. That was how I met Cornelio Reyna's father, who worked at the plant, and Cornelio, himself, who became my friend and compadre." He specifically remembered the many evenings when young men from the neighborhood, including several brickmakers, would gather in front of the pharmacy to chat and sing. Américo first met Cornelio Reyna in 1954, but met Ramón Ayala in 1960 during those informal evening get togethers.

As López described it: "They approached me and we formed little groups in the evening in front of the pharmacy that was open until 1:00 a.m. They gathered there to engage in the life of bohemians [*bohemios*]. They sang among themselves. They played many romantic ballads, and *ranchera* music including a few *corridos*, more than anything songs that carried a message." The one song in particular he remembered them singing was *Te Vas Angel Mío* that had been recorded by *Los Alegres de Terán.* Although Cornelio Reyna worked as a brickmaker alongside his father in Tueme's plant, and elsewhere, López noted that "he did not like that kind of work" and, for that reason, escaped, dedicated more time to playing music, and "got lucky."

On the basis of his experiences during that period, López developed a very positive image of brickmakers and their occupation:

> It was a dignified form of work. They were artisans. According to the preparation and molding of the clay, they put a special stamp on the brick. They were great artisans and people who lived very well. It was rough work but dignified. In those years it was one of the best ways to make a living. The brick plants and PEMEX [*Petroleos Mexicanos*] were best for workers. Seasonally, the cotton gins also provided employment. A majority of houses, around 60 percent, in Reynosa were made of brick but the lesser quality brick sold for a good price and was available instead of wood in working class neighborhoods.

Not unlike agricultural workers who live in proximity to the fields where they labor in relative isolation from wider society, *ladrilleros* enjoyed music and dance as a form of recreation and relaxation after a hard day's work. Accordingly, it is not surprising that budding musicians within their ranks, like Cornelio Reyna and Ramón Ayala, or others living in the vicinity of the *ladrilleras,* like Rubén Vela and his family in Relámpago, through their music improved the quality of life for brick plant workers and their families. Even today in *Ladrillera Reynoso*, Juan Garza can be heard singing *ranchera* or other melodies *a capella* as he molds brick. Another of his co–workers, a brick molder named Lucio Sánchez whose photograph appears on the front cover and as photo 71, works on an irregular schedule at the brick plant because of his success in his second job as a guitar–playing singer of religious songs for an evangelical church.

On the basis of informants specifying songs popular among the *ladrilleros* like *Te Vas Angel Mío*, *La Pajarera*, *Chulas Fronteras del Norte*, and various *corridos*, in other words songs that had a message, it is reasonable to infer that there was a fit between lyrical content and the demands and conditions of brickmaking. This would have been especially true during the period when brickmaking in the Rio Grande Valley was done by undocumented Mexican immigrants. Many of these songs evoke images of itinerancy, surveillance, and/or life's contingencies as experienced by these Mexican workers as they moved from job site to job site up and down and on both sides of the river border. In other words, the association between brick culture and *ranchera* music was probably not coincidental.

Nevertheless, there is no inherent or necessary connection between Mexican brickmaking, or any other type of manual labor, and musical expression. During several years of research in the state of Oaxaca in southern Mexico among workers in several occupations, including brickmaking, I never once encountered evidence of such a connection. In other words, in the case of the border brick industry, we are most likely dealing with a linkage within the *norteño/fronterizo* [northerner/border] variant of Mexican culture that operates across manual occupational cultures including brickmaking. Arguably, music resonates uniquely with working class experiences among *norteños* and *tejanos* in a mutually reinforcing and common way in the "clash of cultural economics and their changing course" (Peña 1999:19–21 et passim). Having said this, however, randomness and personal idiosyncrasies definitely intervene in this process. There is no demonstrable hard–wired linkage running from infrastructure to superstructure, through commodity value circuits, to make *ladrilleros* or other members of the Mexican–descent working class of the lower border musical (cf. Cook 2004: see especially chapters 7 and 8).

Conclusions

A survey of the history of Lower Border brickmaking discloses the importance of two factors: the largely Mexican–origin labor force, and the Texas masonry construction market The 1950s demonstrate the consequence of the induced [by restrictive immigration policy] scarcity in the supply of Mexican labor to brick producers in the Rio Grande Valley: plant shutdowns, increase of the *maquiladora* sector, the relocation of production to Mexico and, in the case of Weiske's Valley Brick and Tile Company, the attempt to make simulated handmade brick with a soft–mud machine. The survival and rejuvenation, albeit on a reduced scale, of handmade production in Reynosa into the first decade of the twenty–first century reflects the persistence of demand for these brick in the Texas masonry construction market together with the capacity and willingness of entrepreneurs and workers in Reynosa to continue to supply it.

Notes

1. Sra. Morales de Garcia read to me from her unpublished *Apuntes* during an interview that was tape recorded and transcribed. The quotation is from my transcription.

2. On Map 2, within the Los Longoria-Calabazas landform [the area in the middle of the map essentially surrounded by a meander in the Rio Grande/Rio Bravo], the location of the *Hacienda Longoria* and the *Ladrillera Reynoso* is indicated by the black dot in the middle of the far left side of the landform, near the river. Many of the other brickmaking sites mentioned in this chapter are located by black dots and white lettering on Map 2.

The simplest English translation of *ejido* is common land, however, this is misleading since it confounds two distinct types of land tenure in Mexico, communal and *ejidal*, that are legally and institutionally separate in Mexico's post–revolutionary agrarian reform system and Agrarian Code [*Código Agrario*]. During the twentieth century *ejido* land was possessed in allotments to individual community members through their elected representatives [*comisariado ejidal*] who served as stewards of land nominally communal but, in reality, part of national patrimony. The land was held in private usufruct with possession/use rights inheritable between generations of family members. The post–1993 reforms facilitated privatization of *ejido* lands.

3. Unfortunately, although the Runyon collection contains more than 8,000 photos, it includes only seven of brick plants—all showing separate views of this same plant.

4. Joe Summers Jr., who worked at and ultimately inherited his father's machine shop in Mission that serviced the Weiske plant over several decades, explained the nature

of extruders and dry–press machines in brickmaking as follows: "Some clays you could run stiff mud, extruded, sheer it off and make brick out of it. And other clays you could not do that. But almost any clay could be constituted to make dry–press. They just press it under high pressure, dry it a little bit, put it into the kiln and fire it. It was solid common brick. It was easier to make them because you could constitute your material and keep your moisture low. There is hardly any shrinkage. It went right out of the press straight to the kiln. The dry–press is the cheapest to operate. You just turn them out, make four or six brick at a time" (1993 Edinburg Roundtable transcript). Commenting on Summer's statement, Ino Weiske observed: "The advantage of the stiff mud extruder is that you can change dies and make different shapes, sizes." And, with regard to the dry–press machine, he also observed: "You just have to have a man with great patience to set it up. In those days, dry–presses were set up with shims. The presses were so designed that they would make each brick the same thickness, or break. The drive belt would slip off the pulley and stop. They could not make an oversize brick" (1993 Edinburg Roundtable transcript).

5. Sources on 'Operation Wetback' include McBride, J., 1963; Acuña 1981:156–162; and Morales and Tamayo–Sanchez 1992:61.

6. According to Peña (1999:45) the popularity of the *tambora de rancho*, a make-shift drum made with wire, rope and goatskin was "fundamentally related to the political economy of subsistence" and was "an instrument attuned to the life–rhythms of an agrarian people."

7. In Spanish, the lyrics doña Minerva Morales could remember are: "*Una morenita enaguas de percál, mi mamá será la arena, mi papá será la cal, mis hermanos los ladrillos, ya esté ladrillado el mar.*"

8. There is consensus among these informants that both Reyna and Ayala worked as *ladrilleros* in Reynosa for a time in the late 1950s/early 1960s. It may be, however, that only Reyna also worked in Relámpago—although Sebastián Reynoso remembered that Ramón Ayala did as well. Since that *ladrillera* closed in 1956, it appears that only Reyna was old enough to have worked there (born ca. 1940; Ayala was born in 1945). It is probable that Reyna and Ayala had work experience in both *Ladrillera Tueme* and *Ladrillera Salinas* which were located in close proximity on the *linea de gas*. When I asked Américo if he knew anything about the brick plant in Relámpago, Texas, he volunteered the following: "When they formed the group, he [Cornelio Reyna] and Ramón [Ayala], that was when they began to talk about working there, in the United States, in that brick plant in Relámpago and from there they brought the name of the Relámpagos del Norte."

There is disagreement in the source materials regarding dating, however. For example, a 2004 interview by Cecilia Ballí for *Texas Monthly* magazine puts Ramón Ayala's birth date as 1945, and has him joining Cornelio Reyna in 1959. Manuel Peña states (1985:106): "Organized in the mid–1960s by singer–bajo sexto player Cornelio Reyna and bajo sexto player–accordionist Ramón Ayala, *Los Relámpagos* was discovered in Reynosa . . . by . . . Paulino Bernal" (cf. Peña 1999:104). Yahoo! Music states: "In 1960, the promising musician [Cornelio Reyna] met Ramón Ayala and decided to team up with him to assemble a norteño act called Los Relampagos del Norte." *Wikipedia*, by contrast states: "Cornelio Reyna formed a partnership with the legendary Ramón Ayala to create the famous duet called *Los Relámpagos del Norte* in 1963."

Chapter Three
The Reynoso Brothers and the Brick Industry in *Ejido Los Longoria*

Commercial brickmaking in the Rio Grande Valley during the twentieth century drew upon the knowledge, skills, and labor of itinerant *ladrilleros* from Reynosa and other riverine pueblos of Tamaulipas. Rio Grande Valley *ladrilleras* that for generations served as training grounds and income sources for Mexican *ladrilleros* also became, in the middle decades of the century, capture points for their deportation by the U.S. Border Patrol (Cook 1998:53–54, 90–93,94, 96–97, 110–111, 140–141). The saga of the Soto family outlined in chapter 2, and Sebastián Reynoso's career of hard work and enterprise to be examined in this chapter, provide a human dimension to these otherwise impersonal processes.

Los Ebanos and Madero in the western riverine portion of Hidalgo county [see Maps 1 and 2] were primary sites for artisanal brickmaking during the early decades of the twentieth century. The major mid–century site was Relámpago in the easternmost riverine portion of Hidalgo County near the Cameron County line. As Map 3 shows, both the Relámpago plant [ABY] and the Toluca Ranch plant [TBY], which closed before the Relámpago plant opened, exemplify the close interface between brickmaking and meandered riverscape characteristic of the lower Rio Grande's deltaic brickmaking belt throughout its history.[1]

Many *ladrilleros* from Reynosa worked first in the Dunn and Solís *ladrilleras* in Santa María, just east of Relámpago in Cameron County, as well as in the Toluca Ranch plant, prior to working in Relámpago. The following recollection by Sebastián Reynoso, who worked at two of these plants as a young man, mentions the participation of master brickmakers like Eduardo Soto, son of Andrés:

> I started working in 1945 in a place called Santa María, in the Mercedes area, on the other side of Relámpago, further toward Matamoros. They had a store there on the highway. The *mayordomo* [foreman] of the brick plant was a Mexican named Eduardo Soto. I worked there from 1945 to 1948, the year of the freeze, when they sprayed water on all of the orange trees, when they all froze, in 1948 more or less.

Nevertheless, it was the Relámpago plant that Reynoso considered to be his finishing school for learning the fine points of brickmaking and the brick business. With this knowledge and saved earnings from Relámpago, Reynoso and some of his colleagues were prepared to play an important role in the accelerated growth of the Reynosa industry after their forced repatriation by the U.S. Border Patrol in the 1950s.[2]

Between 1930 and 1950 the sprawling, 350 square–mile municipality of Reynosa experienced rapid population growth, owing to a combination of migration from the interior of Tamaulipas and other Mexican states, and natural increase. From 1930 to 1950, the city's population grew by a factor of seven and by 1980 it had grown by a factor of forty, a period in which the municipality's population grew by a factor of twenty-four. The city's population in 1980 was 194,693, whereas the municipality had a total of 211,412 (Herrera Pérez 1989:115).

Reynosa's growth was spurred by several factors. First, by employment opportunities in regional agriculture, owing to the completion of the Falcon–Anzaldúas hydraulic projects in the lower basin [*cuenca baja*] of the Río Bravo in the 1950s and the opening of the Bajo Rio Bravo irrigation district. Second, by the discovery of gas fields in the 1940s and the subsequent development of pipelines, infrastructure and processing plants by PEMEX between 1955 and 1966. And, third, after 1965, by the establishment of *maquiladoras* [assembly plants] in accordance with the *Border Industrialization Program* (Herrera Pérez 1989:110–115). At least by the mid–1950s, and even more so during the 1960s and 1970s, another contributing factor to this demographic growth was employment opportunities in the artisanal brick industry.

The take–off of Reynosa's brick industry during the 1950s was a complex market phenomenon involving many variables, factors and conditions. First and foremost, there had to be strong demand for brick. This was, indeed, the case and reflected demographic and economic growth in the city of Reynosa and, especially, in the Rio Grande valley and in off–border cities in Texas like Corpus Christi, San Antonio, and Houston. The demand factor was even more significant than usual in this case because in those years the Texas brick industry experienced increasingly squeezed production capacity that constricted its ability to supply enough brick in a timely fashion to meet surging demand (Cook 1998:Ch.12)

Second, there had to be capital available to finance the establishment, operation and expansion of *ladrilleras* in Reynosa. As it turns out, the financing of the expansion of the brick industry required cooperative pooling from different private sector sources that included landed, mercantile, commercial and trucking investors from Mexico using their own funds or bank loans, and a variety of investors, including brick dealers, truckers, builders, and other businessmen from Texas. One element, above all, stands out from this period: Texas demand determined Texas investment and the latter determined Reynosa production. In other words, the boom in the Reynosa brick industry was as much if not more dependent upon the export market and Texas sources of financing, than upon the domestic market and Mexican financing.

Third, there had to be an adequate labor pool of willing and able workers to enter the brick industry labor market. There was, of course, a relatively small but stable pool of local labor for brickmaking. But the labor market demands of the growth decades required extra–local sources in many other areas of Mexico where life was hard, remuneration low, and brickmaking a familiar occupation. Hundreds of such workers migrated to Reynosa during this period to work in the *ladrilleras*; a significant portion of these were systematically recruited by employers—especially from Torreón, Coahuila and its surrounding area including San Pedro Las Colonias and Mayrán, as well as from the area around Monterrey, Nuevo León including Pesquería. It is important to note that, relatively speaking, during the decades in question the level of remuneration in brick plants was higher than those in agriculture, construction, and the service sector—hence providing a real material incentive for workers to seek jobs in the brick industry. Reynosa's location on the border directly across from the Rio Grande Valley of Texas, and the riverine location of many *ladrilleras*, alone provided sufficient incentives for migrants from the Mexican interior to seek temporary work there as a springboard to illegal entry into the United States.

Fourth, there had to be available land and raw materials to facilitate the establishment of *ladrilleras* and to supply adequate amounts of clay, sand and water to sustain open–ended growth in brick production. Even though Reynosa experienced accelerated population growth between 1950 and 1980, most of that growth was concentrated in the city proper (Herrera Pérez 1989:115). This left many of the outlying areas of the municipality either with agricultural or semi–rural lands that were available for development by the brick industry.

Lastly, not to be discounted as a factor in the growth of brickmaking in Reynosa after 1950 was its easy accessibility to the Texas highways and railroads. This Texas transportation infrastructure was directly accessible through the Hidalgo port of entry, just across the river from downtown Reynosa. Brick produced in Reynosa could be transported on trucks to the freight terminals in Hidalgo for shipment by rail to Corpus, Houston, San Antonio and other Texas urban markets. In fact, this was how most brick produced in the Reynosa gas–fired plants of the 1950s were transported. According to doña Minerva Morales (interview with author 1993): "My husband [Gustavo Garcia] shipped brick by train from Hidalgo, Texas. That little railroad was used a lot, and I mean a lot, to ship brick. The Americans built it early in the twentieth century to ship produce from the Valley. Afterwards the refrigerated trailers came to transport the produce. That is when the train started hauling brick."

Efraín Salinas remembered that his first big client, named Masterson, "shipped brick by railroad cars to Houston," a fact that was confirmed and explained in more detail by Joel Rodríguez who was office and sales manager for *Materiales Salinas* in the 1950s: "I remember that a man named Masterson from Houston was our client. We transported brick on Mexican trucks to Hidalgo. There we loaded them onto railroad cars that transported them to Houston." By 1960 flatbed tractor–trailers had displaced rail freight as the most accessible, reliable and convenient mode of brick transport from the lower border region to the Texas interior.

In short, all of the above factors and conditions were present to facilitate the take–off [*despegue*] and boom [*auge* or *pica*] of Reynosa's brick industry during the 1950s, 1960s and 1970s.

Sebastián Reynoso, the Exemplary Repatriated *Ladrillero*: Migration as a Character Builder

In 1935 Isidra Nungaray and her husband with the surname Lara had a child named Sebastián in Nochistlán, Zacatecas. By the time Sebastián had reached school–age, his family resided in the nearby town of Teocaltiche, Jalisco where they had a business as *torneros* [turners or lathe operators] who turned out wooden objects like tops [*trompos*], yo–yos, and chess sets [*juegos de ajedréz*]. As a free–spirited, adventuresome and independent young boy Sebastián ran away with one of the employees in his family's business, a man who made *molinillos* [wooden hand–beaters to stir hot chocolate], to accompany him on a bus trip to Guadalajara and then to Ciudad Juárez. After leaving Ciudad Juárez, apparently alone, young Sebastián was apprehended by authorities in Torreón who were on the lookout for vagrant children and was sent back by bus to his family.

Not long after this incident, Isidra Nungaray separated from her husband and with her two children, Sebastián and his younger sister Manuela, joined a northward migratory flow in the early 1940s to the cotton fields of the La Laguna region in the state of Coahuila. There, she met Diego Reynoso, an out of work shoemaker [*zapatero*] from León, Guanajuato who had left his family behind to seek a better life in Mexico's border region. Ironically, since cotton–picking provided only seasonal work, Isidra and Diego during the off–season went together to the thriving brickmaking zone in the city of Torreón, the same city where Sebastián's earlier adventure had ended. In Torreón, Diego took up brickmaking for the first time to supplement seasonal employment in agriculture that included stints in the U.S. as a migrant agricultural worker. Now a couple in search of a better life, and attracted by news of job opportunities in Rio Grande Valley agriculture and brickmaking on both sides of the border, Isidra and Diego migrated with her children to Reynosa in 1945.

Sebastián's narrative of his experiences in brick plants in the Rio Grande Valley excluded any mention of his parents or other family members, and gives the impression that he was pretty much on his own from a young age, although he was never homeless nor abused. He mentioned that one of his early trips to the Rio Grande Valley was during a winter cold spell. The godfather [*padrino*] of Sebastián's half–brother Antonio ['Toño'], who accompanied his father Diego [and possibly Sebastián] on river crossings to work in the Rio Grande Valley, recounted to Tony [Toño's son and Sebastián's nephew] that Diego dug a foxhole in the ground to help him keep warm during cold nights—a trick he had learned as a boy to protect himself from gunfire during the days of the Mexican Revolution in his home state of Guanajuato. Sebastián never mentioned to either Toño or Tony whether or not he had shared a foxhole with his stepfather on one of those cold winter nights in the Valley but he probably did.[3]

Possibly owing partly to an idealized image of his biological father and/or in response to the economic hardships he experienced while growing up, Sebastián strongly aspired to achieve material success and upward mobility in his life, and rejected his stepfather Diego as a role model. As fondly remembered by his grandson Tony, Diego was "just a simple working man" who always had time to entertain children and "never had an interest in business or in money." "My grandfather," according to Tony, "only wanted to work to feed himself, drink his beers and simply live his life as a *ladrillero*." This does not describe a lifestyle and orientation that was supportive of Sebastián's aspirations.

This rejection of his stepfather was expressed, as family members attest, by Sebastián never referring to Diego as *papá* and by the fact that Sebastián's children were taught never to address or refer to Diego Reynoso as their *abuelo* [grandfather]. According to his younger brother Toño, who was born in 1953, it was only because Sebastián was labeled '*Húngaro*' [derogatory term in Mexico for a gypsy] by his peers in Reynosa, due to using his mother's rather uncommon surname Nungaray, that when the time came for his obligatory military service at the age of 18 in 1954, he signed–up as Sebastían *Reynoso*—finally adopting his stepfather's surname that remained with him until his death in 1997.[4]

Toño sums up Sebastián's rejection [*rechazo*] of Diego as follows: "He felt no pride in his stepfather because of the different ways they saw things" [*No sentía ningún orgullo por el modo de ver las cosas diferentes de cada uno*]. They worked together but Sebastián always rejected his stepfather as mediocre ["*Siempre tuvo un rechazo hacía su padrasto de mediocre*"]. This was driven home to Toño many times when he was growing up, especially since Sebastián heavily influenced his upbringing and education to make sure that he was properly raised and educated, and would not grow up solely under the influence of his poor laboring and hard–drinking father, Diego.

Sebastián Reynoso's Formative Years as a *Ladrillero* in Reynosa and the Valley

The Reynosos found a place to live in a *barrio* of brickmakers in Reynosa known as *La Chaparral* that, according to Sebastián, had many small brick plants [*había muchas ladrilleras*]. It was located at the intersection of Calle Aldama and Calle Bravo along the river. Toño was told during his childhood that he had already been born when the family lived in *La Chaparral* [*en la chaparral, allí por la Calle Bravo*] and that his father was making brick there. This is only about a block away from the corner of Escobedo and Victoria streets where the Soto family's brick plants were located when Anita was growing up. This area is about eight blocks or so west of the international bridge along the levee [*bordo*], and about two kilometers southeast of the Los Longoria–Calabazas area [see Map 2, center] where both families would subsequently relocate not long after Toño was born when the government condemned the area due to flooding and related flood–control projects, for example, levee construction.

In 1945–46 in La Chaparral Sebastián Nungaray, at the age of 11, began to learn brickmaking from a man named Nieves Aguiñar who, at the time, worked in plants in La Chaparral, and in Santa María in Cameron County, Texas [see Map 1]. As he had done a few years earlier with the *molinillo*–maker in Teocaltiche, Jalisco, Sebastián, when told by Aguiñar that he was going to work in a place called Santa María, said to Aguiñar, "Take me with you."

Aguiñar and his eleven–year–old companion, took a bus from Reynosa to Rio Bravo. Since there was no highway from Rio Bravo to Rio Rico in those years, they traveled by unpaved back roads on a truck, stopping often to cut branches that were bundled up and put on the ground so that the wheels could get enough traction to pass across muddy terrain [*Hacíamos arrollones y poníamos ramas para que el camión pudiera agarrselo*]. Their destination was a place called San Francisco, in the vicinity of Las Flores near Rio Rico, where they crossed the river by *chalán* [improvised raft] and then walked along the railroad tracks until arriving at a brick plant in Santa María [Cameron County, Texas]. In Sebastián's words: "We came to a place that was known then as Las Flores and we crossed over the river there. Then we took the famous 'track' until we arrived at Santa Maria." He worked in Santa María until the plant closed in 1948–1949, at the time of a winter freeze when the orange groves were wet down by the growers but froze anyway [. . . *la helada, cuando se lavaban todos los naranjos, que se acabaron todos*], and then went to work at the nearby Anderson brick plant in Relámpago. Regarding his time in Santa María Sebastián observed: "I dropped in there at the age of ten or eleven and I could not yet work well. I earned very little. Rather I was the errand boy there." [*Yo caí de diez o once años allá, pues, y no podía trabajar bién. Yo ganaba muy poco. Más bién yo era el mandadero de allí*]. [5]

Sebastián remembered that the piecerate for brick molders in 1945–46 was $3.50 per thousand, whereas in Mexico it was six pesos per thousand. Given the dollar to peso exchange rate of 4.80 prevailing at the time, a molder could earn 16.80 pesos per thousand in Santa María or more than double what he could earn in Mexico. The same ratio prevailed into the early 1950s at Relámpago, so the brick molders trek across the river was driven by a material incentive or as Sebastián expressed it simply, "it suited us" [*nos convenía*].

The payment system at the Anderson plant in Relámpago [Map 2 and ABY in Map 4] described by Reynoso was uniquely designed by the foreman Manuel Herrera, an Assembly of God minister, to retain migrant workers who arrived there and also as a form of social control. It was, most notably, a system without money. A 'forced savings' element in this system was apparently requested by workers, even if relatively large amounts of cash paid to them when their period of employment ended presented them with security problems on their return trip to Reynosa or elsewhere in Mexico. According to Reynoso:

> Over there in Relámpago there was no cash, no money [*Allí en el Relámpago, no había lana, no había dinero*]. We arrived there to work and they told us: 'There is no money but we have a store, clothing, whatever else you need, but there is no money for you to pay with' [all transactions at the company store were debited against earnings]. They [plant managers] saw that they were able do that because all the people who arrived there to work said to don Manuel, 'Save the money that I earn for me.' So they adopted a system without money, everyone worked full speed ahead but without money . . . Don Manuel told them, 'The day that you want to leave just tell me so that I can give you your money.' When a person with three or four years of work said to don Manuel 'I am ready to leave', at two dollars a day that amounted to 700 or 800 dollars a year or 3,000 or 4,000 dollars for three or four years. That was a fortune at that time. There were occasions when workers were killed on the way back to Mexico during robbery attempts to steal that money.

Sebastián's vivid, if selective, recollections about Relámpago provide a detailed account of the organization and activities at the plant and suggest that little escaped his attention (Cook 1998:87–88, 93, 96–97). His recollections included the many occasions when, still a teenager, he was picked up on raids by the Border Patrol in the Valley and deported to Mexico. According to him:

> In 1953 wetbacks were no longer accepted in the Valley. They told us, 'If you come back, you will go to jail. Our recommendation is not to come back.' To return was like going to prison. For example, I was deported on a plane to San Luís Potosí, from Harlingen, with what I had on me. There was a squad of soldiers waiting, even a Captain, to take everything away from you. The thing they wanted most was money but they took pocket knives, razors, nailcutters, everything. On the other side, it was the same. They took absolutely everything away from you. What you managed to hide, when you arrived in San Luís they searched you again. After they took your money, they sent you to another government official who asked you where you were from. We all responded that we were from Reynosa and they sent us to Rio Rico. Another time I was deported from Relámpago through Zapata [east of Laredo], so that I couldn't return so easily, and without money. You either had to walk back or depend on someone else who still had some money. That was the motive for the eventual closure of the [Relámpago] brick plant.

Photos 12 and 13 in Appendix 1 show this plant at the time Reynoso worked there. [6]

Repatriation and the Development of *Ladrilleras* in Los Longoria in the 1950s

The Relámpago plant suffered severe attrition in its workforce during 1952 due to intensifying Border Patrol raids and round–ups of undocumented Mexican workers. After experiencing several deportations during that period himself, Sebastián Reynoso by the end of 1952, at the age of 17, decided to withdraw his saved earnings for two years of work at the Relámpago plant. The most direct route from Relámpago to the river was through Rosario Ranch located across the *resaca* or settling basin, and the 'bank' [*banco]* formed by it of the same name, that was only 200 meters or so from the plant itself [see Map 3]. "We arrived back here in Reynosa," remembered Sebastián, "and each of us went to his own house." Regarding the matter of the project to establish a brick plant in Reynosa, he admitted that was the "issue we needed to work out well, where to put a business here" [*allá era donde cuajábamos, donde poner un negocio aquí*]. He then summarized the conversation he had with Santos Castañeda Soto on this matter: Question: How much do you have? Answer: Well, I have so much. [*¿Cuanto llevas? Pues, llevo tanto*] Question: And what are you going to do with that money? Answer: 'Well I don't know, Proposal: 'Let's start a brick plant. You have 2,000 dollars and I have another 2,000. Let's join together' [*Vamos a poner una ladrillera. Tu tienes dos mil dolares y yo tengo otros dos. Nos juntamos*].[7]

Apparently, Reynoso and Castañeda were helped in this process by an even more experienced colleague from the Relámpago plant, 'Beto' Vargas, who was employed as Manuel Herrera's foreman. Reynoso explained Vargas' role as follows: "A person [Vargas] came over [to Reynosa] who by that time had nothing to do with the plant over there [Relámpago], and that person took charge of looking for the land for a *ladrillera* because he was knowledgeable [*conocedor]*, he was a setter, a stoker, he was everything. He looked for the land and it was done." The land was bought, a well was dug, kilns and patios were built and a few workers were hired to make brick in 1953. According to Reynoso, "Santos Castañeda and I began there in Los Longoria where the Balboas are. That's when the fields attached to the brick plant [*los campos agregados a la ladrillera*] began to open. The same people who were employed over there [in Relámpago], well, we were employing them here. That is how the first brick businesses were born here in Mexico."

Several repatriated *ladrilleros* from Relámpago followed Reynoso's precedent but without comparable success. They often failed as independent plant operators because, according to Reynoso, they were drunkards. "They would earn ten pesos and spend twenty pesos on mariachis, women, and the like—squandering their capital." Many of these unsuccessful repatriates ended up working for Reynoso and others in brick plants that were proliferating in Los Longoria and other riverine *colonias* of Reynosa. That fate overtook Reynoso's original partner, Santos Castañeda, who did not last long in the partnership; he was bought out by Reynoso and then continued to make brick at the same plant that Reynoso rented back to him. [*Yo compré el terreno y mi socio lo metí como rentero*].

Some of the first repatriate plants established in 1952–53 maintained ties with the struggling Relámpago plant. Although not yet out of his teens, Sebastián Reynoso evidently had caught the attention and gained the confidence of the Relámpago plant administrator, Manuel Herrera, and a business relationship emerged between them. In late 1952 when Reynoso was at the point of repatriation, Herrera asked him: "Listen, what can we do? We want good quality brick but we don't have enough qualified workers. We have only apprentices. We are making bad products." [*Oye, que harémos? Querémos buena calidad de ladrillos pero no tenemos manos. O sea, tenemos puros aprendices. Estámos haciendo puro producto malhecho*]. Sebastián responded to Herrera that he knew many good brickmakers, and Herrera told him to go and contract some. On the basis of his experience with brickmakers in both in Relámpago and Reynosa, Reynoso knew that many of them came from the Rio San Juan/Pesquería area of Nuevo León near Monterrey, so he traveled there to find workers who would attempt to make their way across the border or who would work in newly established Reynosa plants to make Aztec brand brick for the Relámpago plant. According to Reynoso, through an arrangement made with Herrera and with financing from Herrera to pay workers, for a period in 1953 he made Aztec brick, in molds supplied by Herrera, for transport across the international bridge to Hidalgo and then to Relámpago for firing.[8]

There is, however, another chapter in Sebastián's account of his life in the mid–1950s that was interwoven in his memory bank with his Relámpago experience but apparently occurred after he quit working there and after completing his military service in 1954. During 1955, he worked and lived with a family of truckers, named Canales, in Weslaco, Texas. It is not clear how he met the Canaleses but during his stay with them they offered him advice about how to manage money and how to survive in a binational borderland environment. Given the insecure life that he had experienced as a migrant laborer in the U.S. and Mexico, Sebastián observed that when he came to live with them he "spoke badly about the United States and badly about Mexico." The Canaleses made him see that he "had nothing to do with the United States and nothing to do with Mexico," and that as a borderlander he lived on the margins of both countries.

When Sebastián decided to return to Mexico, the Canaleses gave him a used pick–up truck, worth about $75 at the time, and paid him $360 cash for his work with them. The parting words of the elder Mr. Canales as he handed him the cash were: "Look, this is what you are worth [*Esto es lo único que tu vale*]. Take care of it. Go to Mexico and watch out that they don't rob you because any one of them can screw you [*Te pueden dar en la madre*], and this is the only thing you have. You have to take care of this money because it's the only thing that can carry you forward [*Es el único que te puede sacar adelante*]." Sebastián took this advice to heart, returned to Reynosa in his pick–up truck with cash in his pocket, and immediately started hauling firewood [*leña*] to his own and other brick plants. During this period he applied for a green card [*tarjeta local*] and did business with the Canales family "exporting unfired brick" to them from Reynosa. It turns out that at the Anderson plant in Relámpago and with the Canales family, Sebastián had filed with social security and had what he called "letters of suspension" [*cartas de suspensión*] and of "recommendation" [*cartas de sostenimiento*]. Unfortunately, some of these documents were confiscated by Mexican officials when he legally re–entered Mexico at the end of 1955, and others were lost on one of his previous illegal crossings of the river.

Although Sebastián's activity as a repatriated brickmaker at plants in Los Longoria was interrupted by obligatory military service in the Mexican army during 1954 and his stint with the Canales family in Weslaco in 1955, it expanded rapidly after 1955 through doing business with the Canaleses, with Ramiro Rodríguez of Mission, and with an Anglo–Texan named Acre. Sebastián described the situation as follows: "Ramiro had a tractor trailer and he had his own clients. Acre used to carry a bag of dollars from Matamoros to Piedras Negras to meet payrolls. How much do you want? So much. Let's go. And Ramiro started the same way following Acre's example. Ramiro also carried a bag of money but the money wasn't his, it came from his clients." [*Tenía su trayler y tenía sus propios clientes. Acre traía un moral de dólares desde Matamoros hasta Piedras Negras, entregando rayas. '¿Cuanto quieres? Tanto. Vamonos.' Y así comenzó Ramiro siguiendole el ejemplo de Acre. Ramiro traía un moral que no era de él, era del bolsillo del cliente de él*]. [See Appendix 1. Photo 34 shows the remains of one of the plants started by Sebastián on land he purchased in Los Longoria in the area of the clay pits that are still being excavated by the Reynosos].

In 1958, Sebastián's younger brother, Toño, who was only five years old at the time, remembers that "we were making unfired brick in Los Longoria" that was sold, transported to Hidalgo, Texas, and fired there. The original Reynoso plant, started in 1952–53 with Santos Castañeda, would later become the Lupe Garza plant. Another of Sebastián's early plants in Los Longoria with Castañeda was directly to the northwest of the later Guajardo and Macias plants on land rented from an agriculturist named Estéban Balboa.[9]

Referring to another plant started in the 1960s, that Toño himself would rent in the 1970s, he remembers that Sebastián dug a shallow clay pit that had a perfect mix of raw materials for brickmaking—clay above and a little layer of sand below ["*Hacíanos un barranco de este alto*" (raises hand shoulder high) "*llevaba la liga de la mezcla perfecta para hacer ladrillo, barro arriba y un poquito de arena abajo*"]. At the time, the scattered residents of the area, like the Balboa family, were farmers and dairymen who became brickmakers to cash in on the boom market. According to Toño, owing to Sebastián's initiative in developing brick plants, "agriculture declined here in this area after the brick plants came" [*La agricultura se cayó aquí en la región cuando llegaron ladrilleras*].

The Reynoso System: Brickmaking, Housing, and Politics

Sebastián Reynoso presents an exemplary case of an entrepreneurial artisanal industrialist and land developer who was interested in making money from different forms of land use ranging from extraction of raw materials, brickmaking, real estate development, and refuse dumps. According to Toño, Sebastián was at first reluctant to buy land because he thought it was too expensive. Sebastián, however, said that he bought land in 1952–53 where he partnered with Santos Castañeda in starting his first plants. Subsequently, he bought more parcels of land and rented others, in Los Longoria and in adjacent *barrios* [neighborhoods]. Before the end of the 1950s according to Sebastián, "I began to buy properties and, then, I began to make houses for people."

In the 1960s Sebastián got permission from the government to start opening roads in the area to make land more accessible for development of brick plants, extraction of sand and clay, and housing. He was always on the lookout for available land, with or without pre–existing kilns or housing, where he could put a family or single workers to make brick. If clay or sand were not available at a particular site, he would truck it in from other land he owned or leased. Kilns and rudimentary housing were built as necessary. In any case, rent was paid to Reynoso in the form of fired brick.

During this period Reynoso established a series of shifting, informal relationships with local ranchers/farmers like the Longoria, Pérez, and Balboa families. In the case of Lorenzo 'Lencho' Pérez, Reynoso rented land with pre–existing kilns but subsequently, in the late 1960s, bought land adjacent to Pérez and, at the latter's invitation, provided work for a brickmaker named José Luna, and his two sons, who Pérez had briefly employed but failed to pay. Kilns were built at this site, clay was extracted there, and bricks were made for several years until Sebastián's retirement from the brick industry in 1982. Luna then went to work for Toño and was still working for the Reynosos in 2008 at the age of seventy-four. The clay mill at the new *Ladrillera Reynoso* was built at the site of the clay pit dug by Luna and his sons when they worked for Sebastián. To this day José Luna is grateful to Sebastián for having provided him with a job at a time when he had nothing to eat because Lencho Pérez had refused to pay him for the brick he had made. [See Appendix 1. Photo 67 shows the *batidora* or clay mixer with a dump truck being loaded with clay at the current Reynoso plant. Note in the background the walls of an abandoned clay pit. This was the site of Sebastián Reynoso's earlier plant that employed José Luna.

In comparing his father Toño with his uncle Sebastián, Tony Reynoso believes that his father was always more involved in production than was Sebastián during the peak of his career. Whether or not he had plants on rented or purchased land, according to Tony, Sebastián "never had to personally involve himself in any of the work. He preferred to employ people who would work for him. He did not have to get up in the middle of the night or stay awake all night to fire brick in the kiln. He put people there who would produce fired brick for him. He just checked the quality of the brick being made when necessary." If his uncle, Sebastián, had little love for brickmaking per se, Tony views his father, Toño, as being fond of his craft, as "loving to be involved in plant activities" and "to walk around and observe work that is being performed." He concedes, however, that his uncle did a lot to promote the domestic and export brick business, and that his knowledge of the business was enhanced by extensive travels in Texas and Mexico.

It is evident from Sebastián's own account of his personal and business life that he realized the advantage of having political connections to support his interests as an industrialist and land developer. This is why he became a local political operative with the official status of precinct captain [*jefe de casillas*] for the municipal government controlled by the *Partido Revolucionario Institucional* [Mexico's twentieth century ruling party, PRI or the Institutional Revolutionary Party]. His home in *Colonia Aquíles Serdán* [see Map 4, upper middle section] was regularly used as a voting place in his precinct. According to Tony, Sebastián's political connections gave him access to information regarding land that was being confiscated by the government for tax delinquency, and he began to acquire confiscated properties for less than market value by paying back taxes.[10]

In addition to renting land parcels in Los Longoria and adjacent *barrios* to brickmakers, Sebastián also subdivided some of his land into lots for residential housing and *vecindades* or multi–family apartments with common bathroom facilities. He built these rustic housing units very cheaply from a combination of waste brick and a mixture of lime and sand. This set the stage for an expansion of his business that grew to the point in the 1970s where he rented land to José María Cantú to operate a plant, and had the same relationship with his brother Toño on another parcel. "Here on this side of the levee," Sebastián explained in 1995, "I have fifteen houses, and all told I had thirty houses for the thirty brickmakers that I had. For every brickmaker employed I had a house with water and electric service. The whole purpose of having housing for them was so that they helped me to work." Reynoso's formula for several years, then, was "thirty brickmakers, thirty houses" [*treinta ladrilleros, treinta casas*]. He operated in accordance with this formula until 1982.

I asked Sebastián how he had arrived at this formula. His response emphasized that the number of brickmakers he employed was not the independent variable but, rather, was a function of the carrying capacity of the land he had acquired for the purpose of brickmaking. "I have *cortes* [clay pits plus patios] for thirty brickmakers. I have the *corte* that is the mine [*mina*, *pozo,* or *barranco*], or the source of clay, and I have the *patio* that is part of that parcel of land. I don't sell the clay but, instead, I haul it to the *patio* that is part of the same parcel of land. So, I have thirty *cortes*, thirty *patios*, and thirty *casas*."

Sebastián's nephew, Tony, was particularly thoughtful and articulate in his characterization of his uncle's career:

> He was a natural politician who lived in a space where the brick industry was the way he could combine business and politics. He maneuvered very well among the brickmakers, among the people who were involved in the brick business, not only brickmakers but also *ejidatarios* [and] persons from the private sector. He knew how to operate very well in that context [*ámbito*] but he is not someone who, from my point of view, I would consider to be a brickmaker's brickmaker. Rather, for him, brickmaking was a business where he obtained income; he was a politician of the brick industry and an enthusiast of real estate [*Mi tío era un político de la ladrillera y un apasionado de los bienes raíces*].

Supportive of this characterization is the fact that Sebastián also operated refuse dumps on land parcels with abandoned clay pits.

Off–Border Labor Recruitment

When workers were scarce, and Sebastián had only four or five employees at his main plant, his clients would pressure him to hire more workers. He would do so by traveling into the Mexican interior, often to Aguascalientes. He would recruit workers there by offering them a higher wage and guaranteeing them rent–free housing with free electricity and water. Often, he would return to Reynosa with ten or twelve workers that would increase his weekly output by several trailer loads.

His *modus operandi* on such recruitment trips was to drive into a brickmaking area and start asking workers how much they were earning. If they told him twenty, he would tell them that he would pay them thirty. They would make their own quick calculations and agree to go to work for him so they could earn more money [*ganar un dineral*]. Then they would tell him that they had no money to pay for their travel to Reynosa and no way to get there. Sebastián improvised a solution. He convinced a tractor–trailer driver from Reynosa who had just made a delivery in Agascalientes and was planning to return empty, to take back several families if Reynoso paid their expenses and the trucker's fuel expenses. "I went to get all of them together with their suitcases, bags, everything," he remembered, "whole families or single men, however they showed up." He explained that the brickmaker worked with his wife and children. He made brick, his wife took him lunch, and while he was molding brick she was canting or stacking brick that were in the process of drying. Sebastián viewed this as part and parcel of their contract by piecerate wage [*como si fuera maquila*].

When the recruited families arrived in Reynosa, Sebastián took them to his housing areas and showed them where they would live, and then to the plant where he told them they could start working ["*Pueden empezar a jalar*"]. He noted, laughing, that many of these families would tell him, sooner or later, that they would like to go over to the other side as wetbacks. He would point to the river and tell them, "There's the river, you don't have to ask me for permission to go, you are not tied here, everyone is free here." Inevitably, some families would leave and others would come—but he always tried to maintain a population of thirty workers with families. From Sebastián's perspective, turnover was not always disadvantageous because there were usually several undesirables who had poor work habits, drank too much, got into fights, and abused their wives or children ["*gente que no sabía convivir con su misma familia*"].

The state of Aguascalientes became Sebastián's preferred source for workers. During the boom years, when he found himself shorthanded, he would head for Aguascalientes to places where he knew brickmakers were available for hire, and would offer them a higher wage than they were earning there, plus providing them with transportation to Reynosa, as well as with housing and utilities.

Formalization of Business: From Cash for Payrolls to Checks for Contracted Sales

The main reason why Sebastián Reynoso's first clients as an independent brick producer included bilingual Texas–Mexican trucker–intermediaries like Ramiro Rodríguez was because they enabled him, as a monolingual Spanish–speaker, to do business indirectly with monolingual English–speaking Anglo Texan customers in cities like Houston. This role of Tejano truckers as translator–intermediaries is clear from one of Reynoso's explanations of how his business started to grow:

> So a lot of people started to approach me to work. Why? Because they would go and come, go and come across the border and back, to be with clients over there and that was something I did not do. I went but in a sort of bankrupt way [*Yo iba pero en sistema de quebrar, por decir*], that is, just to see what would happen. But people were telling me to look for truckers who spoke English and Spanish, that is, who were bilingual. Those were the ones that did business because they lived and worked on the U.S. side.

Over the years Sebastián did business with a Houston contractor through a trucker named Lázaro Sáenz of McAllen whose trucks made three or four trips weekly to haul brick from the border to Houston. The owner of the company wanted to do business directly with Sebastián but did not speak Spanish. So Sáenz, who was bilingual, became Reynoso's interpreter, delivered money to him for the company's orders, and then hauled brick back to Houston. This relationship with truckers like Sáenz was an integral part of Reynoso's export business for three decades into the 1980s. It is also true, however, that there were periods when Sebastián dealt directly with Spanish-speaking clients in Texas, received orders, and dispatched brick to the buyer in trucks that he rented on an ad hoc basis.

Occasionally Sebastián's clients would show up and complain that he did not have enough brick on hand to meet a contractual commitment—or that he had set aside brick for himself that was produced from their advance. They would tell Reynoso that all of his inventory was theirs because they had paid for it. He would explain to them that his profit was derived from the brick he could retain in his own inventory above and beyond the amount contracted. In other words, for every trailer load of 12,000 brick sold by contract he added to his inventory an additional 1,200 brick or 10 percent of the amount contracted. So, in effect, the money advanced by the buyer financed the production of 13,200 brick—1,200 of which were retained as a return in kind by Reynoso. As he expressed it: "On every trailer load I earned ten dollars per thousand brick. That amounted to 120 dollars on every trailer load of 12,000 brick. Those 120 dollars that I earned were brick that were set aside in my inventory." This is a more precise description of the type of arrangement described in more general terms by Anita Soto regarding her dealings with Elizardo González and others [see chapter 7].

If his clients continued to press him on the details of this arrangement, Sebastián would reiterate his understanding of their business relationship and then tell them bluntly, "take it or leave it." He considered himself to be a "*comerciante*" [businessman] and "*industrial*" [manufacturer], not a capitalist. Then he would explain to his Texas clients the financial arrangement required to sustain their business relationship: "You bring me money, we do business. You don't bring me money, no business. Your responsibility is as the capitalist partner, and I am the partner that produces. You have to maintain my workers for me. If not, we stop things now." Sebastián called this arrangement "*el sistema de agarrar refacción*." This became his *modus operandi*—he was reluctant to produce brick without a buyer willing to finance production. In retrospect, Sebastián Reynoso's *modus operandi* in the brick business during the boom days of the 1970s gives meaning to the term *sellers market*.[11]

Sebastián also referred to this system as reflecting "dominion over people" [*el dominio a la gente*]. He explained that one needed to pay workers on time, usually by noon on Saturday, so you could show up on Monday and tell a worker to go and round up some more workers. The brickmaker would go tell others that in Reynoso's plant they can count on the payroll being met ["*la raya está segura*"] every Saturday at noon. From Sebastián's perspective, "That is what is important" because "he brought you more people to work." This strategy also reflected the relative shortage of brickmaking labor especially in periods of brisk demand for brick like the 1970s when competition among plants for workers intensified. Without the dominion or sway over potential workers derived from his reputation for meeting a payroll on time, a brick producer like Sebastián would not be able to consistently supply his clients prepaid orders.

This payment arrangement is illustrated through the case of Sebastián's relationship with Lázaro Sáenz. He agreed to sell brick to a Houston contractor, through Saenz as the intermediary, for forty dollars per thousand. The Houston contractor sent Sebastián a check for ten thousand dollars as pre–payment for twenty trailerloads of brick. Sebastián, however, wanted the money deposited in Saenz's account, and for Sáenz to pay cash in dollars when the brick were loaded on his truck. He told Sáenz: "You are the bilingual intermediary. Why are we going to get involved in disagreements? You keep the money. Put it in the bank under your name. You pay me in cash when you come to load the brick. If you do not bring me the cash, the brick won't be loaded." As it turned out this was probably a good idea because one of Sáenz's drivers was picked up by the police on suspicion of involvement in the narcotics trade, resulting in the impoundment of one of Sáenz's trucks. Sáenz came to Sebastián for help in getting his driver and the truck released—a process that would require a considerable sum of money. Sebastián told him: "I am not going to help you. I am the third party. It is your employer that has to help you. Both you and I depend on him. So your problem is just that, your problem."

As someone who had little or no formal schooling and who struggled to overcome illiteracy well into his adult life, Sebastián Reynoso went through a difficult period of trial and error regarding the rules, regulations, accounting, and documentation of export business. He was motivated to conduct business properly and legally, but lacked an orientation as to how to correctly negotiate legal border crossings, and initially had difficulties in doing so. He explained how this led him to consult with customs officers and, finally, to participate in an official training session for untutored Mexican exporters sponsored by U.S. Customs: "When I had problems at the border I went to talk with the chief of U.S. Migration in Hidalgo. Every time I crossed they asked me a bunch of questions. I would try to explain myself well but those officials said I was doing so badly. I would show them my permit for seventy-two hours but they would come down on me hard and ask me when did I cross." Sebastián would explain that he had accompanied a trucker to deliver brick in Houston, leaving Hidalgo at twelve midnight for the eight–hour trip. It would take all day in Houston to deliver brick and complete his business, and he would return to the border within thirty hours from the time that he crossed over from Mexico. But his explanations always seemed to be treated with official skepticism.

During a session with a U.S. Customs official Sebastián was told about the need to present notarized manifests when he crossed the border that contained all the required information regarding the purpose of his visit. At the suggestion of that official, who informed him that the brick business classified him as an industrialist [*industrial*] rather than a businessman [*comerciante*] Sebastián attended an orientation class in which all of the requirements and procedures for importation of Mexican products were explained. In a nutshell, once the notarized manifest or invoice was drawn up, presented when he crossed the border, and entered into the U.S. Customs system, his crossing problems were solved.

Another problem that was resolved through this manifest system was payment. It was customary for payment to be made in cash rather than by check, and payment was made at the time brick were loaded onto the carrier authorized to deliver them, say, to Houston. If payment was not made, Reynoso and other brick producers would resort to confiscating the keys to the tractor trailer and hold them until cash payment was made in dollars. The manifest system required that payment for the indicated quantity of brick at the stipulated value be verified by a photostatic copy of the check issued as payment.

Sebastián's education regarding proper customs procedures, and his new reliance on payment by check rather than payment in cash, occurred just after the incident reported above involving Sáenz. He started supplying brick to *McAllen Materials Company* owned and operated by Bob McAllen. Since McAllen did not speak Spanish, Reynoso dealt with a bilingual employee. He produced brick for McAllen without cash changing hands. "I made the brick, sent it over to him with the invoice, just like I had been instructed to do by U.S. Customs, and he sent me a check. I took a copy of the manifest indicating the delivery of brick with the check from the McAllen company through Customs and they gave me permission to go and deposit the check." Another problem avoided by doing business without cash was to avoid payment in counterfeit bills, a practice that was quite common at that time in cross–border commerce.

Devaluation, Dollarization, and Sebastián Reynoso's Retreat from Brickmaking

The peso devaluation crisis of 1982 sent shockwaves throughout the border brick industry and transformed Reynoso's way of doing business. It occurred precisely at a time that he was already rethinking the logic of earning more of his income from direct involvement in brickmaking as opposed to depending more upon his rental properties. Also, he was forty-seven years old at the time and was starting to think seriously of easing his way into retirement after a lifetime of hard work without much leisure and nine children to support. He explained his thinking about these matters as follows: "I have learned over time that I can earn, let's say, five pesos daily in profit from a brickmaker who is living in my housing or I can earn that same amount in rent from anyone who shows up to rent one of my houses. So I asked myself, what was the sense in operating a brick plant if I could rent all of my houses?" He noted that plants he had previously started on land that he owned were, as of 1993, being rented and operated by others. He attributed his change in orientation to his upbringing in places and to past employers that were very religious and Biblically grounded. Although he himself was not a religious person or a reader of the Bible, he said that the influence of these believers made "something stand out to him" [*algo se me paró*]. He just wanted to live happily, to earn enough to meet his family expenses. He came to the conclusion that "avarice is bad, that the more you earn, the more you work night and day, and you go crazy. So one needs to withdraw from those things." Perhaps, this was Sebastián's way of finally reconciling himself with the memory of his stepfather Diego whose lifestyle he had always rejected as incompatible with a quest for upward mobility.

In any case, although no longer directly involved in the brick business when I interviewed him in 1993, Sebastián still could supply brick to clients through his younger brother, Toño, who was operating the plant he rented from Sebastián. Nevertheless, Sebastián was no longer financially dependent on the brick business but on income from land development and rent. He paid twenty thousand pesos for a multi–hectare tract in the 1970s, had it surveyed and then subdivided into 110 lots that were, in his words, "in possession, without papers" but zoned for "human settlement" [*asentamiento humano*]. His business plan was to sell each lot for a substantial profit on his original investment.

In the late 1970s Sebastián was subcontracted by a Mexican construction company under contract with PEMEX to supply brick for a new 500–unit subdivision in Reynosa. His first deliveries were made at a price of 200 pesos per thousand [roughly nine dollars] at a time when the peso–to–dollar exchange rate had only recently officially shifted from 12.50 to 23. This was in the midst of a period of steady decline in the purchasing power of the peso and of the value of the peso in terms of the dollar that culminated in the official devaluation of the peso in 1982, setting the peso–to–dollar exchange rate at 150. Prior to the devaluation Sebastián was compelled to increase prices per thousand in fifty peso increments, sometimes weekly, until they reached 600 pesos per thousand. In anticipation of another official devaluation, he paid a visit to the engineer in charge of acquisitions and told him that he was going to make his next delivery his last and was planning to pull out of the brick business or, in his words, "Yo *ya no voy a jalar.*" The engineer asked him why, and how he was going to make a living considering that he was a brick producer. Sebastián responded that he could no longer deal with the combination of inflation and devaluation that was making accounting difficult since wages and prices were constantly having to be adjusted upwards. He was anticipating having to double his prices from 600 to

1,200 pesos per thousand and told the engineer, "My workers want me to double their pay, and you are not going to pay double the price for brick." The engineer told Sebastián that if he went out of business he would die of hunger ["*Pues te vas a morir de hambre*"] and Reynoso's witty response was: "You are probably right but that is my occupation" [*Pues, a lo major pero ese es mi oficio*].

In a subsequent conversation with the same engineer, Sebastián was asked to explain the problems confronting him. His explanation was as follows: "I am working and the dollar jumps daily. And prices follow. But I pay my workers in pesos. And I am selling in pesos. I cannot adjust to the devaluation. All the time my heart is palpitating—Za! Za! Za!—because I am not able to make ends meet." The engineer asked what he could do to help and Sebastián told him by transacting business in dollars—that all future sales should be priced and paid in dollars rather than pesos. His proposal was accepted and a contract to that effect was notarized. Sebastián then told his workers that, from that day forward, he would pay them in dollars because otherwise there would be no stability in their relationship due to the wage–price spiral in pesos ["*Todos los días aumento, aumento, aumento, pues, no alcanzamos*"]. Under these conditions, Sebastián was obliged to sell his brick in dollars and, also, pay his workers in dollars.

During the years he was still involved in production, several of the larger, semi–mechanized plants attempted to get Sebastián to work for them on salary as, in effect, a foreman who would produce brick for them with workers who were dependent on him. He never agreed to participate in such an arrangement because he saw no need to do so—he had his own clients who were buying directly from him. "I worked by contract, as an employer in charge of workers and at the same time as a mediator. They brought me money and I was nothing more than a manufacturer, not a capitalist. It is just as if I were a foreman and got half the earnings." Sebastián considered that his *modus operandi* set him apart from other brick producers in Los Longoria who produced brick on a speculative basis or on blind faith that buyers would inevitably appear. For him, being a capitalist was equivalent to being a speculator, and a true industrialist, from his perspective, was not a speculator but was a producer who worked by contract with a particular client, preferably with a builder who paid in advance to have brick made for a targeted number of houses or, as he put it, "the client deposited the money in the bank and then I had to find people to produce brick—this was a sure thing. None of that 'whatever happens, happens'" [*Nada de eso 'pega lo que pega'*].

Antonio 'Toño' Reynoso and the Los Longoria Plants

If Sebastián Reynoso was the exemplary repatriated *ladrillero* of the 1950s, 1960s and 1970s, his younger half–brother, *Toño*, was among the exemplary *ladrilleros* of the 1980s and 1990s. Born in 1953 in Reynosa, Toño's first work experience was in a *ladrillera* in Los Longoria with his father and his brother, Sebastián. It was Sebastián who ran the plant and, according to *Toño*, was his role model: "Sebastián started the brick plant when he was still a teenager. My father never made many brick, he liked to drink a lot, and never made brick as his own business. The one who did it all was my brother."

Toño described in general terms how brick culture is transmitted in brick-making families and relates it to his own case:

> All brickmakers who have kids bring them to help out with simple tasks like stacking and turning over brick in the process of drying. Then comes learning how to use the mold. By the time I was ten or eleven years old I already had a little kiln that I fired several times. I did this alone because they wanted me to know how to do everything so that I could become independent.

Even though he started working in the family brick plant at a young age, Toño was able to obtain six years of formal schooling, and learned to read and write. He regrets not having been able to get additional schooling, and thinks that he would have had more success in business if he had done so.

Apart from knowledge of specific processes and tasks involved in brick-making, Toño learned from Sebastián a strategy of self–sufficiency built around the principle of do–it–yourself whenever feasible. It is his intelligent and innovative application of this strategy, in combination with his deep enculturation in an occupation peculiarly adapted to his community's landscape, that enabled Toño to survive as a *ladrillero* under conditions in which many of his cohorts did not. In his words:

> Here in Los Longoria brickmaking has been around for many years. Many plants have gone under. I have survived in brickmaking because I don't know how to do anything else. I have never worked in anything else other than making brick. I have survived because I make the brick molds [*moldes*], I make the water tanks [*pilas*] for the patios, I make the kilns, I am a mechanic for the trucks, I am the am the plant administrator. All of the above has helped me to minimize expenses. Given the small amount of cash I take for myself, the plant is able to bear it.

Toño's 'jack–of–all–trades' capabilities were apparent during my first visit to his home in 1993 where I found him in the midst of doing mechanical work on one of his trucks.

In response to my questioning about the origins of this economizing, self–help orientation, Toño elaborated:

> One of the factors was that my brother was the same type. Among other things he designed wheelbarrows to cart brick. He was very intelligent. Nowadays, the brickmakers who need molds have to find a carpenter to make them. If they need a rake to use in the kiln or on the patio, they have to go to a blacksmith.

> All of those things my brother looked at as not being difficult to make, and he made them and I learned from him. Everything that I learned was from him. From the age of ten I was driving trucks that my brother had. I also had to help him make molds, and everything else. By the time I was eighteen, I knew how to do everything in the plant.

This intense occupational enculturation and work strategy, from his perspective, left Toño occupationally monocultural—ill–suited for other modes of making a living. But this mono–occupationality has also caused him to realistically adapt to the business cycle, that is, to maintain a steady, manageable scale of output and to avoid over–expanding in good times so that belt–tightening in bad times would be feasible. According to him:

> What I have done in hard times is reduce production and try to take care of money. I have never found it necessary to ask for loans to keep working, to keep making brick. The majority of plant owners mix things up, they take money from the plant to be godparents of a fifteen year old or other things unrelated to the plant. That money is from the plant. As plant administrator whatever money I take out I have to recuperate later.

Toño's prudent business strategy is reflected in a carefully managed work schedule: "I have always worked at the same pace, something like six to eight kilns yearly or the equivalent of 420,000 to 560,000 eight–inch brick. We follow other ideals. My older brothers were raised during the aftermath of the Revolution when there was much hunger. People became accustomed to save money for hard times. We have that mentality of being savers, and not worrying about what others have."

Regarding the land tenure situation in Los Longoria and his own *ladrillera*, that in 1999 was located on a one hectare site, Toño filled in some details:

> At first my brother rented and then he bought land. The majority of people around here rent land. This was an agricultural area. Agriculturalists were not interested in brickmaking. Nevertheless, as owners of the land, they got involved in making brick but didn't find it suitable as a business. So, since they have alternatives, they get involved in raising cattle or planting crops. Here in this region there are ejidos and private property, it's mixed. According to the government this year [1993] *ejidatarios* can, if they want, sell their property.

In 1993, while his brother Sebastián was still alive, Toño paid him rent for the *ladrillera* he was operating but later in the decade purchased it.

Even though the Reynosos also owned an elongated strip of land measuring 33 meters wide by 900 meters in length that has direct river access and includes significant deposits of good quality clay, their strategy over the years has been to conserve it and to buy clay from adjacent landowners whenever possible. Sebastián, however, did operate a plant with kilns right at this clay source for several years which Toño was renting from him in 1993 during my first visit. Sebastián purchased the land for that plant in 1971 and in 1973, at the age of twenty, Toño took over its operation through a rent arrangement with his brother. The two brothers cooperated regarding sales, essentially with the same client base and trucker–intermediaries, so that orders received through Sebastián were supplied from Toño's plant output. By 1999 Toño had stopped producing brick at the original plant, and was operating a new plant on a contiguous two hectare parcel of land that he purchased.

While the practice of buying clay rather than extracting it from his own land seems to contradict Toño's cash–conserving survival strategy, he explained the long–term logic of his approach to the clay supply in 1993 as follows:

> Right now I am buying clay because we have relatively little land. I intend to continue in this business for many more years and we have always tried to buy clay elsewhere so that when, for whatever reason, we are not longer able to do so we will have enough clay on our own land to continue. The cost of clay is minimal. We pay twenty–five thousand pesos ($8.00) per truckload or five cubic meters, enough to make two thousand *adobones*.

Depending upon market conditions, this cost could be passed on to the consumer: "Increases in the price of brick conforms to demand. If demand is high and supply low, then price can be increased, but it may also go down later."

Nevertheless, since Toño had a dump truck and a backhoe, he also supplied clay to other plants extracted from land other than his own. He did so, for example, to Martín Guajardo who operated a *ladrillera* in Los Longoria with two small kilns that held 30,000 *adobones* each. Guajardo had a clay supply on his property but bought *tierra blanca* from Reynoso that must be mixed with clay to produce brick less susceptible to breakage during firing. Toño explained this business relationship as follows: "I bring a truck and people to load it. Then if he [Guajardo] has cash he pays me in cash. If he doesn't have cash he pays me with brick. I take brick and *adobones* as payment. We also sell him firewood at times."

In May 1999 Toño's *ladrillera* had eleven workers, only half of whom were molders who molded an average of 100,000 brick monthly, enough to fill one of this two kilns. This represented roughly ten trailerloads. By this time he had relocated the plant to a new one–hectare site a few hundred meters from its 1993 location, and was embarked on a program of modernization that included new transport equipment like a dump truck and a forklift, as well as buildings. The overall plant site by 1999 was divided into upper and lower sections—the upper section referring to the location of the kilns, *patios*, storage building and office, and the lower section referring to the strip of land going down to the river that has the clay supply where the plant was located previously.

Toño explained the transition as follows:

> My mentality changed a little. My sons have grown up and I see things from another perspective. At times one tends to think too much about the short term and one is fearful. But I moved up here with the idea of expanding and I have expanded. After finishing the building I am doing now [pointed to a storage area and office under construction], I plan to build some roofs [*techos*] because that will be better for me and for my workers. There will be fewer losses due to brick drying in the sun. Brick that dry in the sun get toasted and have more of a tendency to break apart. I have plans to keep working for many more years. There, on the lower part of our land, we have a lot of good clay, so much that I won't be able to use it all in forty years.

As will be described in chapter 9, this planned expansion occurred in partnership with his son, Tony, and, in effect, resulted in a transformed *Ladrillera Reynoso.*

Of all the necessary inputs into the brickmaking process and brick plant viability, labor is the most costly and problematic. Toño emphasized the critical importance of labor force retention to staying in business: "Aside from everything else, one needs to worry about having enough cash to pay workers weekly, week after week. There are plants that lack cash and do not pay their workers. Logically, these workers do not return to work and look for another more reliable employer." During the low market days in 1986, Toño was down to one worker in his plant but production never stopped. When the market picked up, he was able to hire more workers from the area simply because it was common knowledge that he had kept his plant in operation, and paid at least one worker, under bad market conditions. In any case, his experience has been that when market conditions improve workers always show up asking for work. "It is a certainty that when sales go up, so does the number of workers looking for work. People hang around, without work, but show up to work when they see an increase in clients coming to buy brick. That's how I've seen it around here. On the other hand, when business gets bad workers leave on their own."

When I shared with Toño the fact that one of his competitors told me that he was considering embarking on a trip to the Mexican interior to recruit workers, he expressed confidence that most of the workers recruited by his competitor could, if the need arose, be hired away by offering them a better wage and benefits package. Although this tactic has worked for him many times before, he recognized that it did not immunize him from the classic bind of small producers confronted with an expanding market. Expanded output means higher production costs; brick industry workers typically respond to piecerate increases by producing more brick per day at a higher piecerate but will shorten their work week. [See Appendix 1. Photos 35–38 show *Toño* Reynoso and his plant in the 1990s].

The case of Juan Garza Aguilar, born in 1944 in San Luís Potosí, is illustrative. My first formal interview with him was in 1999 when he was working for José Maria Cantú at *Ladrillera Rey–Tam* in Rancho Grande. Juan was born in San Luís Potosí but was raised in Ciudad Miguel Alemán where he worked with his father who was a charcoal maker. Before he took up brickmaking in 1970 he worked for many years as an undocumented immigrant in Texas, including a stint on the King Ranch as a goatherder and fence repairer. After learning to make brick with his uncle in Rancho Grande, Reynosa, Garza worked in many brick plants there and in nearby Los Longoria, including several stints with *Ladrillera Reynoso* where I first met him in the mid–1990s. [See Appendix 1. Photo 40 shows Juan Garza hand–mixing clay at *Ladrillera Rey–Tam* in 1999].

Cantú provided Garza with a place for him and his family to live at the plant site; the house had electricity but no running water. Garza complained about the fact that he had to haul fresh water from a nearby neighborhood. Since he had to mix his clay with water from a well near his patio, Garza's normal output for Cantú was only 700 brick daily for a wage of one hundred pesos [eleven dollars] which, in his words, "maintains us" [three person household in which wife and son also earn income]; he typically worked six days to earn 600 pesos weekly.

At the time I interviewed him, Garza was considering an offer to go back to work for Toño Reynoso who was then in the process of building new housing for workers with electricity and running water. Moreover, Reynoso promised to pay him 150 pesos [$16.70] per thousand, a one–third increase over the Cantú piecerate of 100 pesos [$11.00] per thousand. When he worked for Reynoso a few months earlier, Juan had molded an average of 1,330 brick daily. Reynoso explained to him that at that level of output at the new pay rate, he could earn a full week's salary in only three days. This logic and Reynoso's offer appealed to Juan. As it turned out, Garza accepted Reynoso's offer and, at the age of sixty–seven, was still living and working at *Ladrillera Reynoso* in 2009. Nevertheless, as will be discussed in chapter 9, changes at the new Reynoso plant have also changed how Garza and his colleagues work.

Toño, in the 1990s in contrast to Tano Arriaga, was able to avoid selling unfired brick in periods of low demand, and found that his fired brick, sooner or later, sold. He emphasized that his main problem was maintaining a consistent level of output rather than sales. His present and future success rested, in his view, on his ability to attract and retain workers. He accomplished this by paying relatively high wages on a daily basis, and providing housing with electricity and water for his workers.

In sum, like peasant–artisan basket makers in Oaxaca who Toño learned about in a Mexican movie of a B. Traven story, border brickmolders in a given time period will opt for leisure over work once a target income is achieved (Cook 1995 and 2004:Ch. 2). As he put it: "It's rare that they work longer to earn more money. It's really hard work. Physically it's one of the hardest jobs around. One has to be in good physical condition to stand a full–day's work."

Conclusions

From the perspective of the development of the brick industry in Los Longoria, Sebastián Reynoso was the right person in the right places at the right time. His independent spirit, self–reliance and self–discipline did not prevent him from learning through observation and he was schooled through hard work. His experience, native intelligence and common sense combined to give him an entrepreneurial drive that helped him overcome a lack of formal schooling and illiteracy to succeed in his craft and in conducting business across an international border. His younger half–brother Toño described Sebastián as being "demanding and cold" with his family, "sociable" with everyone else, and very "political." Toño also characterized him as a "man of his word" in business, and acknowledged that he learned most of what he knows about brickmaking and business from his brother—including frugality and skills as a jack of all trades. Sebastián expressed disappointment that his two sons took no interest in the brick business or following his footsteps but recognized with satisfaction that his younger brother had the know–how and desire to do so.[12]

One paradox confronted by all of Reynosa's brick plants in the 1980s and early 1990s was that although there was a steady domestic demand for *adobones,* at a selling price higher than the export price, much of that business was done through government contracts that resulted in deferred payment. Consequently, brick producers preferred export sales. Toño explained the situation: "You can sell cheaper in the U.S. than here but the advantage is that over there they pay. Here in Mexico there is a market but mostly on government projects. And the government is very slow to pay. That takes its toll on you when they delay payment; you get fed up. There is a Mexican company that buys some brick from me but it usually takes them two or three months or even longer to pay the bill. That really takes it out of you. The U.S. market is very good because there's a lot of money there and you don't have to sell on credit."

Staying in business under conditions of a capricious market, and a dwindling supply of labor in a labor–intensive industry, requires an appropriate strategy. Some labor must be retained and, therefore, production must continue whether or not orders are on hand. Hence, inventories are built up to await market upturns. When a market upturn is prolonged and inventories are depleted then pressures build to expand production by hiring more workers. This is what appeared to be occurring in 1999. In a cyclical business of this kind, the key problem is how to retain workers in periods of market slump; the solution often lies in bank credit, drawing upon cash reserves, or deploying current earnings from other businesses. This was the *modus operandi* of the extant Reynosa *ladrilleras* at the end of the twentieth century.

Notes

1. Map 3 is adapted from a 1956 (photorevised 1970) U.S. Geological Survey map for the Progreso, Texas quadrangle. The lettering and location of the Anderson Brick Yard (ABY) and the Toluca Ranch Brick Yard (TRBY) was done by Ray Blanchette, University of Connecticut Publications, for the Border Brick Project under my direction.

2. All interview material involving Spanish–speaking informants presented in this monograph were transcribed by me in unedited form from tape recordings of the entire interview. For this book, these materials have either been translated or paraphrased by me in a way that attempts to convey nuanced comprehension rather than an adherence to rules of strict literality.

3. There were at least two brick plants operating in Santa María during the 1940s, one owned by J.E. Solis, a local storeowner, and another by J.C. Dunn, a local rancher (Cook 1998:65–68). Reynoso did not mention the name of the owner of the plant where he worked but did remember other details: "*No me acuerdo del apellido de los Americanos. Tenían una tienda allí sobre la carretera. El encargado o mayordomo de la ladrillera era un Mexicano que se llamaba Eduardo Soto.*" Perhaps Dunn had a store and Reynoso worked at the Dunn plant, or he worked at the plant of the Tejano, J.E. Solís, who definitely had a store on the military highway.

4. It is likely that Sebastián's first offical use of the Reynoso surname was when he filed for Social Security as an employee at the brick plant in Relámpago, Texas during the early 1950s.

5. There were two bad freezes in 1949 and 1951 that wiped out citrus groves, including those belonging to Harry Anderson (Cook 1998:75, 83). In 1949, Sebastián Reynoso was 14 years old. Since the Relámpago plant was operating by 1946 with gas–firing kilns as Har–Lan Clay Products, it is likely that the Reynosos worked there before 1950 when Harry P. Anderson became owner. It is possible that Manuel Herrera also worked at the plant for its pre–1950 owners C.L. Landrum and W.E. Hardwick (Cook 1998:82–83).

6. Sebastián Reynoso, like all of my informants, referred to himself and others who worked across the river in Texas without documentation from immigration authorities as *mojado*—a term that literally means 'wet' but is typically translated as a short form of the phrase *espalda mojada* or *wetback.* For them the term is not considered to be perjorative, insulting, or an ethnic slur, although it is obviously so in everyday discourse in the Rio Grande Valley, throughout Texas, and, for that matter, in the rest of the United States. Unless I am quoting or paraphrasing my informants, I will refer to *undocumented* status regarding border crossing as just that.

7. Santos Casteñeda Soto was the son of the daughter of Anita Soto's father's eldest brother, Policarpio (see Cook 1998:43–44). After his stint with Reynoso, he worked mostly in the Soto family's plants.

8. Despite its official 1956 closure date, Anderson's Relámpago plant by the start of 1953 was no longer molding brick there and was importing both fired and unfired brick from Mexico. Owing to the increasing number of raids at the plant and deportations of his workers, Manuel Herrera began hauling two truckloads of brick daily from Mexico to Relámpago over the new bridge at Progreso that opened in December 1953. A loaded truck was parked on the Mexican side for an early morning crossing, proceeded to Relampago to unload, and then returned to Mexico to reload and cross the bridge again before its 5 p.m. closing time for customs processing. The truck would again unload at Relámpago and then return empty to Mexico around 11 p.m. to start the process again the

next morning (Cook 1998:90–91). At first unfired brick made in Mexico were hauled to Relámpago for firing there. By 1955, however, all *Aztec* brand brick were molded and fired in Mexico—mainly in the area between Nuevo Progreso and Rio Bravo.

The information in the above paragraph comes from my 1995 interview with William C. Cain of Cain Custom Brokers, Inc. whose father built and operated the bridge at Progreso. He was the bookkeeper and part–time toll collector for his father's bridge company.

9. One key historical source makes the following observations regarding deportation efforts during 'Operation Wetback': "In the fiscal year 1953 the INS deported 875,000 Mexicans . . . In 1954 it deported 1,035,282, in 1955 256,290, and in 1956 90,122. The accuracy of the figures is questionable since the INS added estimates of the number it assumed were scared out of the country to the number it actually apprehended and deported" (Acuña 1981:157). That the INS campaign was effective in the Rio Grande Valley is suggested by the permanent shutdown of the Relámpago plant in 1956. Nevertheless, many of my informants were in agreement that Border Patrol raids in brick plants in the Rio Grande Valley occurred earlier than the official launching of 'Operation Wetback.' In my interview with Sebastián Reynoso, he mentioned more than once that problems with the Border Patrol in Relámpago began in 1952 and asserted "En '52 fué cuando ya no aceptaron mojados en todo lo que es el valle." He also specified that "*En el mismo '52 comenzé yo a abrir la ladrillera aquí en Mexico con los mismos clientes. Yo hacia ladrillo 'Azteca' que era la marca de Relámpago.*" In a separate interview, Santos Castañeda also said that he stopped working at the Relámpago plant in 1952 due to Border Patrol raids (Cook 1998:91). Bryan Finley who held summer jobs at Anderson's Olmito plant in 1949 and 1950 remembered that Border Patrol raids were a constant problem even then (Cook 1998:91).

10. Wikipedia defines 'precinct captain' as "the individual who acts as a direct link between the party machine and the voters in the community. The precinct captain helps with voter registrations, meeting new residents of the area or neighborhood, and helping voters get to the voting booths or precincts." This is precisely the role Sebastián Reynoso played for decades as a PRI official in Los Longoria and adjacent areas in the municipality of Reynosa.

11. The term *refacción* carries a meaning from the days of *haciendas* in Mexico where it referred to an expense incurred to sustain or keep the *hacienda* in operation ["*gasto que ocasiona una hacienda para su sostenimiento o impulso*"] [*Diccionario Porrúa de la Lengua Española* 2002:640–641 Mexico, D.F.: Editorial Porrúa]. I analyzed the relationship between the wage fund and the circuit of capital, using Sebastián Reynoso's *modus operandi* as a case in point, in my 1998 book (pp. 205–206).

12. Sebastián also had 7 daughters. He died in 2005.

Chapter Four
The Gas Line and the Capitalist Development of the Reynosa Brick Industry

The importance of natural gas and gas–firing technology to the take–off and expansion of Reynosa's brick industry was highlighted several decades' after the fact in a lead article in El Mañana, Reynosa's major newspaper, of November 3, 1988 headlined ***BRICK INDUSTRY WILL DISAPPEAR*** **[DESAPARECERA INDUSTRIA DEL LADRILLO]**. The caption under an accompanying photo of Gustavo García Garza, president of the Association of Brick Producers of Reynosa [*Asociación de Fabricantes de Ladrillo de Reynosa*] and proprietor of *Ladrillos Reynosa, S. de R.L.*, reveals that the industry's demise would be a likely consequence of a policy decision by PEMEX, the parastate enterprise *Petroleos Mexicanos*, to stop supplying natural gas to industries, such as brickmaking, that did not use it as a raw material for petrochemical products. According to García Garza only three of the nine brick plants belonging to his Association were still producing but at 40 percent of capacity, amounting to 150,000 brick per week. This compared to 500,000 brick per week produced by all nine plants in 1985. The cutoff of their natural gas supply, according to him, would represent a *tiro de gracia* [death blow] to Reynosa's brick export industry.[1]

It was the construction of the first natural gas pipeline [*linea de gas* or *gasoducto*] through the Reynosa area in 1950 that determined the location of the export *ladrilleras* which were able to connect to this line to obtain all of the relatively low–cost gas they needed to fire their kilns. Map 4 includes a tracing of the location and trajectory of the natural gas pipeline in Reynosa [on the map look for the broken line identified as *gasoducto*].[2]

Favorable market conditions for building materials in the early 1950s attracted new entrepreneurs to Reynosa like Efraín Salinas Benavides, who was the first to take advantage of the local PEMEX infrastructure to employ natural gas to fire brick in his new plant. During the 1950s other entrepreneurs, like Juan Tueme, would follow Salinas's example in building new brick plants employing gas–firing technology.

The Gas Line, the Producers' Association, and An Industry Divided

As a recently graduated civil engineer from the University of Nuevo León in Monterrey, Salinas came to Reynosa and started a construction materials business. Not long thereafter he also started a cement block factory and a tile factory. In 1952 PEMEX approved his application for gas service and he established the first *ladrillera* in Reynosa with gas–fired kilns. In a few years this plant grew from a one kiln to a six kiln operation; each kiln had a 180,000 brick capacity. According to him: "The market came to me immediately because there was a lot of demand . . . There was a boom in brick export . . . others also opened various brick plants." His maximum level of output at this plant over the years, with between fifty and sixty workers, was about one million brick per month.

For the original *ladrillera*, Salinas acquired land through what he called a "futures" operation [*comprar a futuro*]; he paid rent with the understanding that after three years he would buy, that is, a "rent with an option to buy arrangement." He did, in fact, buy the land in an area that is today completely urbanized. According to him when he bought it, "it was on the outskirts of the city, land covered with brush that I had to clear. There were no paved roads or anything. The gas line passed by there, I made an application, and gas service was installed." As discussed in chapter 2, Salinas took the initiative to visit and consult with Guenther Weiske at his Madero, Texas plant regarding the technology of gas–firing, and succeeded in getting Weiske to assist with the installation of such a system at his Reynosa plant.

Although he made money over the years by producing and selling brick, he emphasized that, in the final analysis, he may have made more money indirectly, that is, through his ownership of the land where his brick plants were located. For example, he paid no more than one dollar per square meter in the early 1950s for his original plant site measuring approximately fourteen thousand square meters. When I interviewed him in at the plant site in 1992 he estimated that the land was worth between two hundred thousand [$65] and three hundred pesos [$100] per square meter!

Like most of those who followed on his heels in establishing *ladrilleras* along the original PEMEX natural gas line in the 1950s, business success together with the rapid urbanization of Reynosa, caused them to outgrow their original plant sites, and to have to expand into the periurban zone comprised by Reynosa's dependent outlying *rancherías* and *ejidos*. Salinas expanded his operations in two separate riverine areas in the 1960s, *Ejido Rancho Grande* and *Ejido Los Longoria*. The latter site was simply a supplementary plant with four wood burning kilns and ten to twelve workers; it was in operation from the early 1960s until the early 1980s. In *Ejido Rancho Grande*, on the other hand, he acquired a ranch where in the mid–1960s he started to produce unfired brick which were trucked to the main plant for gas–firing.

The man in charge of that operation, Tiburcio Rodríguez, began working for Salinas in 1953 when he was thirteen years old. Around 1960 he started working at the site in *Ejido Rancho Grande* when Salinas offered him fifty-five pesos [$4.50] per thousand to make unfired brick there which would be trucked to the main plant for gas–firing in one of its six kilns [one hundred thousand capacity each]. Since this was twenty-five pesos [$2.00] per thousand above the main plant rate, he accepted the offer and moved to Salinas' ranch. Within a short time, Tiburcio had hired twelve brickmakers to make brick for him. He paid them forty pesos per thousand, paid five pesos per thousand to have the brick hauled to the main plant, and pocketed ten pesos per thousand profit. This ranch operation produced an average of fifty thousand to sixty thousand unfired brick weekly over a period of several years. By the early 1980s the ranch brick plant, run by Tiburcio, had twenty-four workers, and a new gas line had been installed through a cost–sharing arrangement with two other plant owners, Juan Tueme and Magín Villaseñor, who also constructed new plants along the gas line. This installation made possible the construction of four new gas–firing kilns with a sliding protective roof. With between twenty-five and thirty thousand square meters, this ranch plant was double the size of the original plant.

Salinas' multi–site enterprise was typical of the Reynosa–area brick industry during the boom period. Most of the industry's major players, and a multitude of minor ones, display such an involvement which in several cases is compounded by one player replacing another at the same site, either through a sale or rental arrangement. Perhaps the most prominent case of the latter was Magín Villaseñor's sale of his Cavazos plant to Juan José Varela which Varela further expanded as *Ladrillera Industrializada Rio Bravo.*

Another complicating factor was that many of the lesser–capitalized entrepreneurs in the brick industry, like Magín Villaseñor and José María Cantú, were often victimized by the boom and bust cycle in the brick export market. Bankruptcy of one enterprise and subsequent reorganization, which meant a new name and often new plant sites, was common in their cases and many others. It was also common for Reynosa–based entrepreneurs to buy or lease sites or plants elsewhere in the riverine zone. For example, Juan José Varela had another plant in Rio Bravo; Raúl Valera had plants in Miguel Alemán and Rio Bravo; for a year and a half in 1979–80 José María Cantú leased the *Ladrillera Ramírez* in Ciudad Díaz Ordáz; and Gustavo García acquired the *Ladrillerías Camargo* plant in Camargo [see Map 1]. In short, multi–site involvement and changing tenure and use, combined with the proliferation of brick plants of all sizes, as well as satellite fields throughout the riverine zone over a period of several decades into the 1980s, make it a formidable project to accurately reconstruct the spatial record of brickmaking and its impact on the landscape. This is true despite the fact that some of these plant sites and facilities, inactive for two decades, remain *in situ* as visible symbols of deindustrialization.

As a pioneer in gas–fired brick plants in the Reynosa area, Salinas' experiences vis–à–vis mechanization merit special consideration. His first step in mechanization was introducing a simple dry pan or roller mill to grind clay and a pug mill or clay mixer so that his brickmakers would no longer have to mix their clay manually with hoes. His intention was to lower costs by increasing worker productivity and to improve quality. Instead of purchasing a new or used pug mill or mixer [*amasadora* or *batidora*], in the early 1960s Salinas built an improvised system out of used materials and components—a strategy widely employed in the border brick industry. In his words: "Even though pug mills were available on the market, I made one by improvising. I made a pug mill and supplied ready-mixed clay to the molders. I was the first locally to use a pug mill."

For Salinas, as was the case with all of the largest–scale brick plants in Reynosa's riverine zone, the process of mechanization stopped short of the actual brickmaking and firing process. The bottom line was to produce a handmade brick that could favorably compete with machine–made brick in the Texas market. He explained: "The point was to lower costs in handmade brick . . . I always had the idea that a handmade brick that would meet U.S. specifications and would withstand competition from machine–made brick."

Regarding the issue of pursuing mechanization to the point of displacing handmade with machine–made brick he was equally unequivocal: "I always was of a mind to mechanize the different processes in handmade brick production without changing the type of brick, not to use a mechanical brick press. Mill clay mechanically, yes. Mix clay mechanically, yes. Transport mixed clay to the molders, yes. But making of brick by hand, don't change this."

In 1959–1960 a crisis emerged in the Reynosa brick industry related to complaints lodged with the Mexican government by representatives of the regional building industry regarding the scarcity of brick and cement in Mexico. Although it is not clear if, in fact, the scarcity was more characteristic of cement than of brick, it is true that all of the largest brick plants in Reynosa at the time, like *Ladrillera Gómez* and *Materiales Salinas*, were producing exclusively for the export market. In any case, the complainants succeeded in getting a temporary embargo placed on brick exports.

At the time this crisis erupted, Alberto Gómez already had more than a decade of experience in producing brick in Reynosa's riverine zone for export to Texas. According to Joél Rodríguez, referring to the early 1950s when he ran the office for *Materiales Salinas*, *Ladrillera Gómez* was the first brick export plant in Reynosa. Like *Materiales Salinas*, *Ladrillera Gómez* produced brick entirely for export, but, apparently, its kilns were never converted from wood–burning to gas–firing.

Sebastian Reynoso recalled that an engineer representing the firm of *Constructora Mexicana, S.A.* complained to him during a visit to his plant in the late 1950s that preference should be given by Reynosa's *ladrilleras* to Mexican rather than to U.S. clients. Reynoso explained to the complaining builder that the sale of Mexican brick to U.S. clients "is not a matter of races, racism or anything like that; it's a matter of who pays best." Referring to the example of a particular buyer from Houston who had cornered the market at that time Reynoso scolded the engineer: "That one [buyer from Houston] pays one-hundred dollars per thousand, you pay sixty. He pays in advance. I transport the material to you and you tell me to come back Saturday for payment, and Saturday, there's no money, and none on Monday or Tuesday, either. Not until the following Saturday do I get paid. That is the reason I do business with him."

The Mexican construction industry sought to remedy the alleged shortage of cement and brick by seeking the intervention of the *Secretaría de Comercio* [Department of Commerce] in Mexico City. This resulted in a temporary suspension of the export to the United States of these products that meant, in practical terms, that trucks hauling Mexican brick were not allowed to cross the Reynosa–Hidalgo international bridge.

To counter this move by builders, Reynosa's export brick producers had to form an association to represent their collective interests by seeking to reverse the federal government's temporary trade embargo. This ad hoc association of brick producers was named the Riverine Zone Brickmakers Association [*Asociación de Ladrilleros de la Zona Ribereña*] and included large gas–firing ladrilleras like *Materiales Salinas* together with wood–firing plants like *Ladrillera Gómez* and *Ladrillera Reynoso*. Alberto Gómez served as the Association's first president. Gómez was involved in producing handmade brick for export as a contemporary of Andrés Soto and Daniel Valera but in the early 1950s was also a pioneer in producing machine–pressed brick. According to Raúl Valera: "In 1953–54 I had brickmaking machinery installed. There was already one other producer, Beto Gomez, who had machinery. He began a little earlier than I did. His was the first mechanized plant here; then I mechanized."

Thanks to the Association's lobbying efforts, the brick embargo was lifted but the Mexican government established new export licensing regulations, since the brick industry was exploiting a nonrenewable national resource and with the intent of alleviating the alleged shortage of brick for the domestic building market. The result of this government policy was to promote the development of a two–tiered brick industry—what Reynoso referred to as "*la grande industria*" [big industry] and "*la baja industria*" [low industry]—one for export, the other for the domestic market. This boiled down to a division between the smaller, unmechanized, wood–burning enterprises and the larger, semi–mechanized, gas–burning ones. As he put it: "Then the rich guys came into the picture, Chato Salinas, the Arab Tueme, and other guys with money. Their attitude was that they didn't want to sell domestically but only to the United States." Reynoso, as a small producer, was cynical regarding this division that he referred to as "the system as always in which the rich get all the power and the rest of us are marginalized. They are jockeying for power and have others marginalized because it's in their interest to do so" [*Aquellos están jineteando, y a los otros los tienen marginados porque les conviene*].

Alberto Gómez asked Reynoso to undertake a survey of brickmaking in Reynosa for the Association. Reynoso claimed that the survey identified 230 *ladrilleras* of all sizes in the area that had a dependent population of seven thousand workers and their families. He clarified that the survey did not refer to seven thousand *ladrilleros* but to seven thousand "mouths, including children, adolescents, and grandparents." He described it as "a list we compiled to present to the Department of Commerce in Mexico City so that they would give us an official identity." These numbers suggest that whereas the largest *ladrilleras* were export–oriented in 1960, there was still a plethora of small *ladrilleras* producing a great deal of brick for the domestic and export markets.

The Department of Commerce sent a team of investigators to Reynosa that visited brick plants to conduct surveys. According to Reynoso: "They showed up at a plant and began to ask questions to workers: What do you eat? What do you drink? How do you dress? How much do you earn? And then to us, the producers: How do you pay? What are your selling prices? How much do you earn? Everything was written down." Apparently, this investigation resulted in a suggested pricing structure for the industry that, in Reynoso's words "was fair, people earned and we earned so that we could all live well" but, he added, "we could not raise or lower prices, they were fixed by the central government." This government attempt at price stabilization in the border brick industry was short–lived.

Joel Rodríguez's recollection of the motive for establishing the *Asociación de Ladrilleros* coincides with Reynoso's: "The Mexican government closed the door to exportation and a permit had to be gotten to export brick. This was because brick became scarce in large cities like Monterrey. All brick was being exported so the needs of our country were not being met. That is what motivated us to hold a meeting." He elaborated: "Through the Association we approached the Mexican government. Blanket permits were processed and then a distribution was made supposedly 'equitable', among each plant. That's how we controlled ourselves."

Magín Villaseñor expressed the view that the Mexican government's closure of the international bridge in Reynosa to brick traffic was unfounded since the real problem in the Mexican construction market in the 1950s was a shortage of cement in Monterrey created by a huge export demand from the United States. According to him, there was not a shortage of brick for the domestic construction industry. His recollection was that the purpose of the Association's lobbying of the federal government in Mexico City was to expose this error and to convince the government to remove the embargo on brick which was based on incorrect market data.

As Rodríguez implied in the statement quoted above the division of the market had built–in inequities among Association members, making the process equitable among them was difficult. Moreover, most of the smaller, wood–firing brick plants, that he refers to as "people of another system" [*gente de otro sistema*], were not members of the Association. Sebastian Reynoso, for example, became involved in a domestically oriented effort that he referred to as "the group for maintenance of the nation or of the city" [*grupo de mantenimiento de la nación o de la ciudad*] where he and other small producers agreed to supply the needs of the local construction market, including public housing projects. However, he viewed this as "having to make promises that I could not keep" since domestic prices for brick were too low to allow for a profit. He told one government contractor that he wanted to meet his commitment to them but would have to hang himself if he did so ["*Yo quiero quedar bien con el compromiso que nosotros hicimos con el gobierno y ahora me tengo que ahorcar porque si no me ahorco, ¿como cumplo contigo?*"]. So, in the last analysis, Reynoso did not comply with the program, and was insulted for being a "sell out" [*vendido*].

In any case, by 1966 the export market was reopened to all brick producers due to a change in Mexico's export permit policy that abandoned blanket licensing through business associations for a standardized ad hoc documentation process. This had the effect of limiting the Association's ability to influence the brick export market, and strengthened the hand of truckers who not only hauled brick for the plants but, increasingly, were also becoming independent or freelance suppliers of brick for Texas customers.

In 1933 Daniél Valera, born in Reynosa in 1880, took up residence on Aguascalientes Street in *Colonia Rodríguez* of Reynosa, and shortly thereafter started a *ladrillera* which exported handmade brick to Hidalgo where it had been used in masonry construction for decades. In 1954 Daniel's son Raúl, born in 1920, established his own mechanized export plant in Reynosa where he produced ten thousand twelve–inch long brick called *barras* [bars] daily with a dry–press machine and, also, ten thousand handmade brick daily. According to him, only Alberto Gómez preceded him in Reynosa with the production of machine–made brick. His plant's output grew to forty thousand brick of each type daily with a total of about three hundred employees. Valera recalled that "there were only a few workers on the machine" and that he "hired people by the day but on contract. I told them to make forty thousand brick and then to go home. I bought the machine used from a brick company in Macon, Georgia." Although Raul Valera was one of the first to produce machine–made brick in Reynosa, the bulk of his export production of between two million and three million pieces monthly was handmade at his plants in Reynosa, Miguel Alemán, Rio Bravo, and Nuevo Progreso. According to Valera, handmade was cheaper to produce in Mexico than machine–made, and was more favored by clients in the Texas and Louisiana markets where he sold.

Valera was forced to shutdown his mechanized Reynosa plant by 1960 due to "a very serious problem with unions." He claimed that he was forced to shutdown by the union unjustly because his workers were earning twelve pesos [$1.00] daily that was more than the official minimum wage at the time. It was precisely to avoid union difficulties that he established new plants in Rio Bravo and Nuevo Progreso. Efraín Salinas gave considerable weight to this factor as a rationale for the establishment of a producers association: "We were fifteen or twenty producers and the idea was a regional border association that covered Piedras Negras to Matamoros. We united in order to resolve the problems that confronted us. We had very serious labor problems. We had a strike."

In the 1960s the leadership of the producers' Association transferred to Gustavo García Garza and its name was changed to *Asociación Regional de Productores de Ladrillo.* In 1974, after orchestrating a successful defense against unionization of the Reynosa–based gas–firing plants, it became legally constituted as an *asociación civil* and became officially registered as the *Asociación Regional de Fabricantes De Ladrillo, A.C.* Gustavo García remained as President until his death in 1991. In the 1970s and 1980s the Association made ad hoc interventions on behalf of producers to resolve problems pertaining to trucking and customs regulations, but was not active in the realm of marketing or in addressing quality control problems of artisanal brick entering the export market during those years.

The 1974 constitutive act establishing the *Asociación* states its general purpose as "defending the interests of the brick industry" or "through association develop interests of collective benefit in both the material and moral sense." These "interests of collective benefit" were specified as: 1) Defense of the interests of members and their collaborators; 2) Cooperate with the effort to obtain positive results in the manufacture of products; 3) Promote friendly relationships among associates; 4) Promote relationships with similar local and foreign groupings to achieve the goal of strengthening the brick industry, and to stimulate the exchange of ideas about common problems; and 5) Cooperate with charitable and cultural institutions. The nine signers of this act were: Gustavo García Garza, Eleazar Varela Ramírez, Efraín Salinas Benavides, Ramon J. Longoria Gaitán, Eduardo Reyna Creado, Magín Villaseñor Hinojosa, Juan José Varela Herrera, Raúl Valera Garza, and Elizardo González González.

Although the newly constituted Association, like its predecessor, purportedly covered a jurisdiction including the municipios of Nuevo Laredo, Guerrero, Mier, Díaz Ordáz, Reynosa, Rio Bravo, Valle Hermoso y Matamoros, or the entire "northern border of the state of Tamaulipas," its membership was much reduced in number to nine from nineteen previously, and in municipality representation [notably missing Nuevo Laredo, Ciudad Miguel Alemán and Ciudad Díaz Ordáz]. Therefore, it was less regional than its predecessor and its membership was mostly representative of the export–oriented Reynosa–based gas–firing plants. Also, noteworthy is the exclusion of non–Mexican enterprises and individuals from membership in the Association (Art. 10, claúsula d, pp. 3–4), as well as the stipulation that each member must "have his enterprise registered with the corresponding authorities and with the National Chamber of Industries of Transformation [*CANACINTRA*]" (Art. 10, claúsula e, p. 4). Relevant in this regard is the fact that Gustavo García Garza, a major plant owner in the brick industry, was the first President of *CANACINTRA* in Reynosa, a position he occupied at the time the new Association was constituted.

Masonry Construction, Brickmaking, and Two Ramones in Reynosa and Camargo

Problems in the domestic brick market were instrumental in drawing Ramón Longoria and Ramón Márquez into joint ventures in brick production. Their early partnership in construction projects in Reynosa carried over into artisanal brickmaking in Reynosa and, later, into artisanal and mechanized brickmaking in Camargo.

Ramón Márquez, born in the state of Jalisco in 1930, started working in ladrilleras in Gómez Palacios when he was eight years old. The going rate of pay was fifty cents per thousand brick molded. He worked for four years in Gómez Palacios and then moved to Monclova where, at the age of twelve he made and fired brick on a portion of his family's residence lot [*solar*]. He remembered the kilns as being small and that they cut a type of grass called *gobernadora* to use as fuel for firing. He also worked as a mason's helper [*peón de albañil*] in Monclova prior to arriving in Reynosa in 1946 at the age of sixteen where his elder brother was already working in construction.

According to Márquez, when he arrived in Reynosa the city was small and jobs were hard to find. but he and his brother made a go of it, working a few days a week and spending a lot of time fishing in the river. But, starting around 1948, job opportunities improved and, from then on he and his brother had as much construction work as they could handle, starting with the new PEMEX building and then with home construction. His brother became a contractor and got the contracts, and then Ramón worked for him as a sub–contractor. They worked together like this for several years, using mostly handmade brick in their projects.

From 1949 to 1954 the Márquez brothers operated a plant in Corrales where they made *ladrillo* and *adobón* for their own construction projects, and for sale. Other plants at that time were making *adobón* only on a special order [*encargo*] basis. According to Márquez, *adobón* "was cheaper from the standpoint of construction labor cost because with smaller brick two were necessary for the width of the wall whereas only one *adobón* was required." In those years, in Corrales there were only about eight brick plants averaging four or five workers each. By contrast, Las Calabazas and Los Longoria each had about fifteen brick plants with small kilns ranging in capacity from fifteen thousand to twenty thousand pieces.

In 1955, Márquez began working full–time as a mason [*albañil*] in the PEMEX construction project, and in 1960 through his involvement in the construction business he met Ramón Longoria. He built homes for Longoria in *Colonia Ribereña* [Map 4, upper center] on a *turnkey* basis but not without some friction between them. The problem was budgetary: Longoria told Márquez to make him a cost estimate for the project but was critical of Márquez for charging as much as an engineer. Márquez told Longoria that the only difference between himself and an engineer was that he *works* and the engineer does not, and that he could not work for less because work is work. Their business differences were finally ironed out, and Márquez built five houses for Longoria, working on a per diem basis. Longoria was pleased with his work and told Márquez that he would continue to work with him but on a contractual, rather than a per diem, basis. Márquez earned about twenty-four thousand to twenty-five thousand pesos [$2,000] per completely finished house.

The building boom in 1950s–1960s Reynosa, illustrated by the large *Colonia Ribereña* [Map 4, upper center] project of Longoria and Márquez, was accompanied by an undersupply of brick. Echoing the earlier statement by Márquez, Longoria remembered that "when I was building in Reynosa I needed more brick. There was not sufficient *adobón* or large brick for my projects." This led him to a meeting with Magín Villaseñor, who Longoria described as the "most famous brickmaker in Reynosa who made good brick," to work out a deal for *adobones*. Villaseñor said that it was not cost–effective for him to produce *adobones*, so Longoria then offered to buy an interest in Villaseñor's enterprise. A partnership emerged between them and the *adobón* supply problem was remedied. After forming their partnership, Longoria and Villaseñor saw an announcement in Reynosa newspaper that a brick plant owned by an Italian named Grimaldi was for sale in Camargo. The Grimaldi plant, and another owned by the Sáenz family, made Camargo an important center of regional brick production.

Longoria recalled: "Over a cup of coffee, Magín and I talked about the Grimaldi plant. We took over the plant on a lease–buy basis. With the rent Grimaldi paid off his debts then we refinanced with the bank. That's how it was done, we bought the plant. We gave money to Grimaldi so that he could travel to Italy for two years." The buy–out process was completed in June, 1967 and in September, as fortune would have it, Hurricane Beulah hit the region and destroyed much of their multi–million peso investment in the plant.

Longoria's partnership with Villaseñor in the Camargo *ladrilleras* lasted only thirteen months during which time, according to Villaseñor, "We made the plants very large." The relationship between Longoria and Márquez, on the other hand, was longer lasting, if also somewhat strained. Márquez's memories of the period parallel those of Longoria: "Brick of the type we used [one and one–half inches thicker than the normal of two inches] was scarce because there was a lot of construction." They were buying brick for their projects in *La Ribereña* and another subdivision from the Grimaldi plant in Camargo but the supply was insufficient for their needs. According to Márquez, this undersupply situation is what caused Longoria to become involved in Camargo and, eventually, to mechanize the plant there.

During this period Longoria focused his attention on mechanizing the plant acquired from Grimaldi, and Márquez was in charge of expanding handmade production, including the building of several large molding and drying sheds and related facilities to the north of the Camargo plant toward the river. According to Longoria, he and Márquez had around 500 employees in their brickmaking operations in Camargo that included the mechanized plant and the satellite operation for molding and drying handmade brick. It was necessary to supplement mechanized with handmade production in those years because the output of machine–made brick was not sufficient to fill the available kiln capacity.

Hurricane Beulah in 1967, which included a hail storm [*granizazo*], essentially destroyed the unmechanized plant, together with 800,000 unfired brick. In the aftermath of that disaster, Márquez became a twenty percent co–owner, with Longoria, of the Camargo plant [*Ladrillera Mecanizada Ribereña, S.A.*], and became supervisor of the plant's mechanized production. Still, in a 1970 letter from the Association to the Port Director, U.S. Customs, Hidalgo, Texas announcing "prices on hand–made soft–mud brick . . . f.o.b. their respective plants in Mexico," three additional plants in Camargo [exporting via the Rio Grande City, Texas port of entry] are listed either under the proprietorship of Longoria [*Ladrilleria Ribereña, S.A.* and *Ladrillera El Alamo, S.A.*] or Márquez [*Ladrillera La Enseñada*].

Joél Rodríguez Flores and the Transformation of *Ladrillera Reynosa*

In 1950, at the age of fourteen, Joél Rodríguez Flores started working as a sales clerk and office boy at *Materiales Salinas*, a building materials supply business in Reynosa owned by Efraín Salinas. When Salinas established a *ladrillera* in 1952, Rodríguez worked with him there in sales and accounting. In his words: "I was the one in the office, the one who sold, and bought. The engineer and I were the only ones there. I did the list of workers' salaries. I paid, sold, and collected. It was a small business: the engineer, myself and the mayordomo, Gilberto Benavides. After a time the business grew. We had a secretary and I worked exclusively in sales." During those years, Rodríguez also managed to study for a degree in accounting at night school.

By 1960, having learned a great deal about the brick business at *Materiales Salinas*, Rodríguez accepted an offer from a Mr. Andrews of Hidalgo, Texas to partner in the start–up a brick plant in Rio Bravo. Andrews invested capital and paid for the *mayordomo*, and Rodríguez managed the plant for a fifty percent share of profits. The plant had four kilns with a total capacity of one hundred thousand brick but, after only one year in operation, it failed due to a combination of factors. According to Rodríguez: "I was unable to set prices. The market set prices. We did not have roofed sheds and the rains affected us badly. Costs rose steadily." Rodríguez learned that when a product is sold for a given price that it has to be produced at a cost less than the selling price. He was unable to produce brick in sufficient quantities on that basis. This experience taught him that it is one thing to work for an enterprise as an employee and quite another to be owner. He thought that his experience at *Materiales Salinas* would have been enough to enable him to be successful with his own brick plant but found that this was not the case. As he expressed it: "I thought I knew what was required to run a business, but I did not know enough."

After this initial failure as an independent businessman, Rodríguez worked for two years as an accountant for an oil–well drilling company sub–contracted by PEMEX. However, he had been bitten by the brick bug during his earlier experience and by the desire to prove that he could run a business: "I wanted to prove to myself that I could make money at making brick. I sought to understand why I had failed. Did I lack capital? Did I lack other resources? Did I lack administrative experience?" During this period of employment and self–examination, he was able to sell his remaining brick inventory in Rio Bravo through a business contact in Dallas, Texas.

It was not long until Rodríguez was presented with an opportunity to find out if he now had the right stuff to successfully operate a brick plant. He was asked by Gustavo González, Carlos Solís, and Gustavo García to participate in a partnership organized to integrate separate brick plants in a common marketing arrangement. Originally, these plants were located near those of Salinas and Juan Tueme. René Ramirez, who started working at this original plant in 1961, remembers that they had no machinery, fired with gas, had three kilns of forty thousand to fifty thousand capacity, and produced eighty thousand to one hundred thousand brick monthly for export. Rodríguez explained his partnership arrangement as follows:

> We formed a company of sorts. It was curious because although it was a company we operated independently, we were partners in sales, we had a common sales operation, but each partner had his independent plant. The original partners had an opportunity to buy land where the plant is now located on the riverine highway in Rancho Grande and they asked me to become a partner in the land acquisition. Then we formed the company with the understanding that each one of us would build four kilns each and would start producing brick. Each of us produced independently and had an assigned output. When sales came, I sold one week, another partner sold the following week, and so on. That is how we produced and sold brick in the company.

Missing from Rodriguez's account of this early history is that a primary reason for relocating the *ladrillera* to Rancho Grande was to avoid unionization. According to René Ramírez, plant supervisor: "There were problems with brickmakers who wanted to form a union. The four partners had seen the land at this location and saw the possibility of establishing a new brick plant there. So they shut down their other plant."

The new company still consisted of four separate entities operating through a market sharing arrangement but with no coordination in quality control and pricing. Through a combination of mutual discussion and outside advice, the partners eventually decided to form a single enterprise with one brand name, *Ladrillera Reynosa* or *Reynosa Brick Company,* with Joel Rodríguez as general manager.

The ownership composition of *Ladrillera Reynosa* underwent further changes through the death of one partner, Solís, and the voluntary withdrawal of another, González. Finally, in 1977 the remaining two partners, Joél Rodríguez and Gustavo García, amicably agreed to divide the plant between them as described in a letter [November 3, 1977] to the Administrator of U.S. Customs at the Hidalgo, Texas port of entry: "This plant known as *Ladrillera Reynosa S. de R.L,* located 3 kilometers from the city of Reynosa on the Riverine Highway and whose proprietors Gustavo García Garza and Joel Rodríguez Flores, as of the first of the present month was divided as follows: On the south side *Ladrillera Reynosa, S. de R.L.* remained with Joel Rodríguez Flores, separated by the gas line; on the North side, *Ladrillos Reynosa,* with the same quality of brick, became the property of Gustavo Garcia." Rodríguez agreed to pay a royalty [*regalía*] to García in order to retain the name of the original enterprise.[3]

A physical visit to this site [or a 'virtual' visit by satellite via Google Earth] reveals that its layout has changed little since the ownership division was made in 1977. On the northern side of the site, a somewhat decrepit long-roofed structure is part of *Ladrillos Reynosa* belonging to Gustavo García's family; and, across the road to the south, Joél Rodríguez's imposing *Ladrillera Reynosa* encompasses three hectares with a series of five molding and drying sheds [*casilleres or secaderos*], each measuring one hundred feet in width by two hundred feet in length to give sufficient space for fifteen separate patios. To the west of the last molding shed is another roofed structure, measuring two hundred feet wide and one hundred feet deep, that houses the clay grinding and mixing machinery. There are four units of three kilns each located to the north of the molding/drying sheds. Between shed one and two is a roofed structure that serves as a workshop. Between the two easternmost kiln units is a wide, open semi–trailer loading area. The main office is located in the northeast corner of the plant site.

René Ramírez, long–time plant supervisor at *Ladrillera Reynosa*, was born in Ciudad Miguel Alemán in 1938 and at ten years of age began working in a brick plant belonging to Guillermo Ramírez Acosta, only one of two plants in Miguel Alemán at the time, the other belonging to Benito Barrera. Brick were handmade. There was no brickmaking machinery, no pug mill [*batidora*], or clay digging equipment. Clay was prepared with hoes [*asadones*]. Ramirez had fifteen to twenty workers. René's stepfather hauled water to mix the clay, while he helped to stack [*torrear*] and cant [*cantear*] unfired brick. His stepfather was foreman and he taught René how to fire the kilns at the plant which were both gas– and wood–fired. Most of the thirty thousand to forty thousand brick produced weekly were exported.

After his stint in Miguel Alemán, René went to work in Camargo at the Grimaldi plant. Grimaldi's plant also fired with gas but did not yet have roofed sheds for molding and drying. It was about the same size as Ramírez Acosta's plant in Miguel Alemán. At the age of twenty-two, after working for a year in Camargo, he came to Reynosa. He and his stepfather made unfired brick along the river and sold it to *Ladrillera Mexicana* [*LAMEX*]. Their intention was to operate their own plant but they lacked the necessary start–up capital. Raúl Valera asked them to operate a plant for him in the riverine zone. He paid them a certain amount of money and sent trucks to pick up a few thousand brick weekly. They looked for landowners who were willing to rent land for brick plants, rented the land, and hired ten or so workers. They dug clay, prepared it for molding, molded and dried it, and then trucked it to the larger plants on weekends for firing. That was in 1959–1960.

Regarding the labor force in those years, Ramírez said that lots of workers were available and would show up at the plants on their own. There were many brick plants in the riverine zone, in Calabazas, Los Longoria, and Rancho Grande. Many landowners rented land for brick plants so the entire area was saturated. So, there was no need to go looking for workers. Those who showed up looking for work already knew what to do—they did not require training. Retention was difficult, however. Workers would show up, work for three weeks or so, and then move on. René elaborated:

> Workers would come from Torreón or from the Monterrey area. They would show up at the plant, ask for money for food since they were always in bad shape. They would work for two or three weeks and then leave. We would lend money to some who would work for two or three days and then disappear. That is one of the reasons we went broke and had to shutdown the plant. Mostly they were single men. We provided them with provisional shelter at the plant.

Given the improvised nature of the plant he operated with his stepfather in Reynosa, Ramírez candidly admitted that it was not really a business since it would rain overnight and ruin the brick that were covered with discarded cement bags since plastic sheeting or tarps were too costly or not available. Much of their output was ruined and this left them with no money to pay workers. Given the problems encountered in trying to make a go of producing only unfired brick, René and his stepfather went to work in the *ladrillera* of Gustavo González [that evolved into *Ladrillera Reynosa*]. At the time René went to work for González, the *ladrillera* had three other co–owners, one of whom was Joél Rodríguez. As foreman, René earned a fixed salary of sixty-five pesos [$5.20] weekly, only five pesos more than the average molder.

According to Ramírez, the four co–owners produced brick separately and had different policies for their workers: some provided on site rent–free housing as did Juan Tueme and Efraín Salinas in their plants; whereas others, including González, did not. At that time, their workers had to rent housing in the surrounding area. After Joél Rodríguez became a partner, he constructed permanent housing on one edge of the plant property for workers and their families. Despite the expense, this proved to be a sound investment for plant owners since the availability of rent–free housing and utilities provided a strong incentive for workers to become permanent, and also tended to improve productivity since unwaged household labor helped out the wage earners. In Rene's words: "The worker with a family, well, they help out by stacking and canting brick, and other tasks; whereas the single worker has to pay children to do those tasks."

During the 1970s boom years at *Ladrillera Reynosa*, Ramírez remembered that workers became scarce and it became necessary to recruit them from outside the region. As sales expanded and production grew, he noted, "We began to search for more people to make brick, to increase output. Trucks were sent to *Pesquería* near Monterrey to bring workers to the border." At this time, every available patio in the plant's five molding and drying sheds was occupied and unfired brick were also being produced outside the plant. All twelve gas–firing kilns were in operation, nine with capacities of around 90,000 brick each and three larger kilns with capacities of around 180,000 each.

Conclusions

The 1950s growth spurt in Reynosa's brick industry reflected surging demand from the domestic construction market as well as from the Texas construction market. In the 1960s and 1970s runaway growth in the industry was spurred mostly by the Texas building boom and by squeezed Texas brick manufacturers who were unable to supply that surging market. By all accounts this was a hectic period of wheeling and dealing, of stunning success stories and equally shocking failures, of huge fortunes made and large debts incurred, of some enduring partnerships formed but even more that were fleeting and dissolved in bitterness and recrimination. Aside from a myriad of stories and testimonials in the memory banks of those who participated, the most striking legacy from this period remains in the *municipio's* landscape—especially in the ejidos of Corrales, El Banco, Calabazas, Rancho Grande, and Cavazos blighted with decaying remains of abandoned *ladrilleras*, many now transformed into trash dumps or simply vacant lots— and many marginalized families and neighborhoods [Map 2].

This entire saga is, perhaps, most striking for the tapestry–like complexity of the social, economic and cultural relationships it involved. Men and women of different regional, family, and class origins and backgrounds in Mexico—from a variety of places in Tamaulipas and many other states in northern and central Mexico including Veracruz, Nuevo León, Coahuila, San Luís Potosí, Guanajuato, and Zacatecas—as well as Anglo–Texans and Mexican–Americans from the Rio Grande Valley and off–border regions of Texas.

Notes

1. The entire matter of the natural gas system and the supply of natural gas to users in the *Zona Geografica del Norte de Tamaulipas* (Northern Geographic Zone of Tamaulipas) that includes Reynosa, Matamoros, Rio Bravo and Valle Hermoso was finally resolved by the Mexican federal government in Resolution number RES/154/97 of 1997 that, in effect, allowed the remaining members of Gustavo Garcia's Association to continue to be supplied with natural gas. As of that date there were only two surviving brick companies that did so: Garcia's *Ladrillos Reynosa*, run by his widow Minerva Morales de Garcia, and Joél Rodriguez's *Ladrillera Reynosa.*

2. Map 4 is a digital scan done by Holland Photo Imaging (Austin, Texas) for the Border Prick Project of a portion of an official map of Reynosa (G14B83) produced and printed by the *Dirección General De Geografía Del Territorio Nacional*, San Antonio Abad No. 124, México 8, D.F. in 1985 (2nd printing). The position of the international border in the map corresponds to the official blueprint of the International Boundary and Water Commission dated 24 November, 1970.

3. This plant site is located by a black dot on the lower left side of Map 2 just above the white lettering for *Rancho Grande.*

S. de R.L. means *Sociedad de Responsabilidad Limitada*, a form of limited stock company in which partners are liable only up to their contributions. More common is an *S.A.* or *Sociedad Anónima* which is another form of stock company with limited liability requiring a minimum of two shareholders and a higher minimum share capital that the S. de R.L. It has a board of directors and foreigners may sit on the board. In 1977, when Joel Rodríguez and Gustavo García split the original company, Rodriguez's new company retained the original name *Ladrillera Reynosa* but the corporate structure changed from S. de R. L. to S.A. de C.V. (*Sociedad Anómina de Capital Variable*) or a variable capital company. Unlike a regular S.A., the S.A. de C.V. does not require a change of its charter if there is a change in its capital.

Chapter Five
Brick Fever, the Periurban *Ejidos*, and the Field Boss System

The metaphorical expression "brick fever" [*fiebre del ladrillo*] comes from my interview with Jesús Ruíz Sánchez [b. 1942, Reynosa] who spent much of his life working as a kiln setter [*asentador*] and foreman [*mayordomo*] in brick plants owned by Magin Villaseñor, Emeterio Salinas [Efraín's brother], and Juan José Varela. For him, the expression evoked specific conditions in the late 1970s, when he was working in *Ladrillera Industrializada Rio Bravo* [Cavazos] for Juan José Varela. "When one says brick fever," explained Jesús, "one refers to . . . a lot of work in brick, everywhere. The brick plants sold all the brick they made. People were working in brick plants all over the place. At the Varela plant, for example, we worked shifts from 6:00 a.m. to 1:00 a.m. the next day." Jesus lived with his wife and nine children in *Colonia Juárez* in the city, even though he was offered a place to live in Cavazos by the Varelas. A bus from the brick plant picked him up every morning for the half hour ride to the plant. No sooner did they arrive at the plant than he and other city–dwelling workers were hard at work. There was so much work that he would come to the plant early on Sunday mornings and by noon had set 25,000 brick in a kiln. In his words: "And no sooner did we arrive than we got to it, we worked as hard as we could. I was thinking that the job was never going to end. Then suddenly, damn it, nothing!" [See Appendix 1. Photo 14 shows Jesus Ruíz demonstrating how brick are set in a kiln].[1]

How did brick fever originate, intensify, and reach its peak? What were the identities of its protagonists? Where and how did they participate in industrial activities generated by it? Answers to these questions will be explored in the remaining sections of this chapter and in chapter 6.

The Periurban *Ejidos* and Brickmaking

Reynosa's *ejidos* have provided conditions congenial to the development of the artisanal brick industry. The municipality's expanding urban core is bounded, especially in its riverine areas, by a series of *ejidos* that collectively encompass hundreds of hectares of essentially rural land. It is precisely in these riverine areas of the municipality where the combination of alluvial and sandy soils are most suited to brickmaking: Corrales, El Banco, Las Calabazas, Los Longoria, Rancho Grande, and Cavazos [see Map 2, white lettering]. The mixed land tenure and use status of these areas—*ejidos* with interspersed private properties where agriculture, ranching, and brickmaking are widespread—make them resistant to urbanization. Even so, the *ejido* lands in Los Longoria have, over the years, been reduced through government expropriations related to the need for public housing or flood control. Nevertheless, the tenure regularization policies of the 1990s, while promoting the change of *ejidatario* usufruct into private ownership, have not yet eliminated rural land use and settlement patterns.[2]

As Map 2 shows, the riverine areas of Reynosa with most relevance to artisanal brickmaking, El Banco and Los Longoria–Las Calabazas, are very much shaped by the peculiar meanderings of the Rio Bravo. As its name implies El Banco itself sits on the bank of an oxbow lake [*resaca*], a truncated section of a meander in the river channel that forms a 'U' shape. This body of water is now a settling basin [*estero*] and is polluted by "*aguas negras*" [water contaminated by sewage]. Los Longoria–Las Calabazas is located on a peninsula–like, misshapen landform bounded on the west, north, and east by the meandering Rio Bravo, and connected to Reynosa [and Mexico] on the south where the distance between the river's east and west meanders narrows to about one-half kilometer.

The southern boundary of Los Longoria–Las Calabazas can be represented by a line drawn from the easternmost point of the *Rincón Bend* on the river to the westernmost point of the *Pate Bend* [see Map 5]. If the river were to be re–channeled along this half–kilometer line the entire territory of Los Longoria–Las Calabazas would be north of the international boundary. Indeed, such a re-channeling or 'cut–off' of *bancos* occurred in the 1950s, not only reducing the size of this area but also eliminating *ladrilleras* that were in operation at the time [see Maps 4 and 5]. Guadalupe Garza Balboa was among the *ladrilleros* forced to relocate by the rechanneling which he referred to as "when they cut–off the river" [*cuando se trozaban el río*]. His family had twenty hectares of ranch land in the area, and today the Balboa ranches remain private property. In short, the unique riverine political geography of Los Longoria–Las Calabazas makes it especially vulnerable to flooding and to natural and man–made changes in the course of the river.[3]

Ejido Los Longoria was established as a grant [*dotación*] of 717 hectares plus a *parcela escolar* to thirty heads of household [*capacitados*], approved by a presidential resolution dated October 22, 1941 (*Expediente Num. 26, archivo de la Procuraduría Agraria, Reynosa, Tamaulipas*). This grant consisted of various tracts separated by private properties contiguous to the population of the city of Reynosa (*Doc. Exp. Agrario de 23 enero 1983 "Investigación general de usufructo . . ."*). The *hacienda Longoria* was partially expropriated and divided into a combination of *ejido* allotments [*parcelas*] and private properties [*pequeñas propiedades*], at least one of which was owned by Alfonso Longoria [and his son, Reynaldo] who in the 1950s and 1960s became facilitators, through land and credit, of the Soto family *ladrillera* in this area.

As of 1982, according to an official "Definitive Socioeconomic Study of the Ejido" [*Estudio socioeconómico del ejido definitivo* in the *Expediente Agrario*], the original grant had already been reduced to 390 hectares owing to an expropriation by CORETT [*Comité para la Regularización de la Tenencia de la Tierra*] and INDECO [*Instituto Nacional para el Desarrollo de la Comunidad y de la Vivienda Popular*] of 127 hectares unsuited for agriculture and earmarked for public housing. The specific purpose of this expropriation was described as the regularization of land tenure in several *ejido* residential areas that invaded *ejido* land, and it resulted in an indemnization. Also, *Recursos Hidraúlicos* [federal *Department of Hydraulic Resources*] expropriated thirty hectares for a project to straighten the course of the Rio Bravo without indemnization. The lands lost to these two expropriations were classified as "irrigated land" [*tierra de riego*] and represented a substantial loss to *ejido* agriculture. The 390 hectares of *ejido* land remaining after these two expropriations consisted of 100 hectares of land irrigated by pump [*de riego de bomba*], thirty hectares of seasonal land [*de temporal*], and 260 hectares of pasture or grazing land [*de agostadero*]. [4]

The INDECO expropriation was compensated through an indemnity [*indemnización*] of 9,525,000 pesos [$423,333] that, in turn, financed two important land purchases by the Los Longoria *ejido*: a parcel of 180 hectares of irrigable land in the municipality of Rio Bravo and a three thousand hectare cattle ranch in Rancho Grande with eight-hundred head of cattle. These two acquisitions increased the size of the *ejido* to 3,570 hectares, larger than that of the original grant, apportioned as follows: 2,960 hectares pasture land, 280 hectares of irrigated land, and 330 hectares of seasonal or rainfall dependent land. Given its greatly increased access to agricultural land the ejido acquired two large equipped tractors with a 5,500,000 peso [$245,000] loan from *Banco Longoria, S.A.* The 1982 study sums up the situation as follows: "It can be considered that 90 percent of all the land of the *ejido* are suited to agriculture or cattle grazing," and further noted that the *ejido* had seven equipped tractors.

Documents in the *Expediente Agrario* indicate that Los Longoria was originally a "collective *ejido*" but that the newly acquired land would be divided up into 120 hectare parcels with the *ejidatario* having the right to fence land that could be cleared for agricultural purposes. The remainder of the land was designated for "collective cattle–raising" with the provision that *ejidatarios* had permission to cut trees for fence posts and for firewood.

Ever since the *ejido* acquired lands through indemnity payments, these have been targeted for takeover attempts by "new population centers" under the pretext of abandonment or non–use by the Los Longoria *ejidatarios*. These attempts have been countered by the *ejidatarios'* defense that "from their date of acquisition . . . these lands have been exploited agriculturally and for cattle–raising through the collective *ejido* regime by the residents of *Ejido Los Longoria*."

In 1985 the land base of the *ejido's* original grant was once again reduced through the expropriation of 55 hectares by PEMEX to install gas wells and collectors. The indemnization report related to this expropriation noted that the expropriated land was 60 percent grazing land for cattle and 40 percent seasonal agricultural land planted in corn and sorghum. The land was valued at eighty thousand pesos [$180] per hectare, resulting in a payment to the *ejido* of 1,975,639 pesos [$4,390]. It is not clear from the documents examined in the *Expediente Agrario* exactly how many *ejidatarios* were impacted by these changes, however, it can be inferred that the number was less than thirty. The 1983 "General Investigation of Landholder Usufruct" [*Investigación general de usufructo parcelario . . .*] refers to "the small portions of land that still remain in the *ejido* from the Presidential grant" as including "ten parcels that are not exploited in any way collectively" and others exploited by nine *ejidatarios*, plus the school parcel" [not counting family members]. The average size of the cultivated parcels [either in corn or sorghum] is nine hectares, and in several cases "the land is worked collectively by two *ejidatarios*."

The lack of documentation of brickmaking in the *Expediente Agrario* is somewhat surprising, given the significant role this industry has had in the ejido's economic life since the 1950s. This lack probably reflects the reality that brickmaking has mostly been practiced by non–resident, non–*ejidatarios* through informal rent agreements with individual *ejidatarios*. For the most part, such agreements were made with the knowledge and tacit approval of the local *comisariado ejidal*. Through such agreements, non–resident renters recruited workers, built kilns, and established *ladrilleras*, with and without the active participation of the *ejidatarios* from whom they rented the land. Over the years, the *ladrillera* workers became *avecindados* [settlers] many of whom themselves subsequently became own account [*por su cuenta*] *ladrilleros* who were also renters.

The *Federal Agrarian Code* [*Código Agrario*] prevailing prior to the 1990s, incorporated a mechanism allowing for productive relationships between *ejidatarios* and outsiders known as "agreements of association and participation" [*convenios de asociación y participación*]. Ing. Patricio Garza, head of the Agrarian Reform office in Reynosa in 1993, explained: "We, the Agrarian Reform Delegation, through the Secretariat, make agreements [in which] the producer of anything, in this case brick, rents land in order to produce the material. This is a legal agreement, following the laws, and is one of the programs offered by the Secretariat of Agrarian Reform to benefit the *ejido* community: agreements of association and participation. Much of the brick production in Reynosa's *ejidos* was undertaken, either formally or informally [i.e., without official documentation], in accordance with these program guidelines."

Some *ejidatarios* did become directly involved in operating *ladrilleras*. One example is Gilberto Arredondo of *Ejido Calabazas* who produced ladrillos for about five years beginning in 1983. According to him, he was the last *ejidatario* to become involved in brickmaking. He worked outside of the *ejido* for many years as a ranch administrator, but decided to become involved in brickmaking as a supplement to ranching and farming. As he explains: "We had to look for another way to make a living because we had only farming and ranching but a lot of family. We have ten hectares of land for planting but so many in the family that we had to find other sources of income to live better." He acknowledged that Sebastian Reynoso provided financing and helped him to operate his brick plant but he also produced brick, with their molds and brand plates or frogs, for Anita Soto, Juan Tueme, Gustavo García, and José María Cantú. Regarding these arrangements, Arredondo disclosed: "Every week they gave us money to pay the people working for us . . . they had money and clients on the other side who supported them in paying to have brick made. If that help is not there, there is no brickmaking." As an *ejidatario*, he had the right to cut and haul firewood from areas of the *ejido* that had *monte* [uncleared brush], and also to dig for clay which he did the old fashioned way with hoes and shovels.

Arredondo's son worked with him along with two or three hired hands in these tasks and also in brickmaking. Arredondo did not mold brick himself but did work as an *asentador* setting unfired brick into the kiln. He had to hire a stoker [*quemador*] to fire his kilns but he also hauled his unfired brick to other *ladrilleras* for firing. Even though he entered the brick market at the tail end of the boom, Arredondo said that during those years many "*gringos*" [one of whom was African–American] from Houston drove their tractor trailers into his ranch and visited and ate at his house.

It would be mistaken to infer from Arredondo's account that he and other *ejidatarios* were as hard–pressed economically as were many of the non–*ejidatario avecindados* who rented land from *ejidatarios* to make brick. In terms of housing and household furnishings, other assets like vehicles and farm equipment, animals, and land, *ejidatarios* like Arredondo were relatively affluent in comparison with the typical non–*ejidatario* who rented his brick plant. Arredondo's emphasis on requiring funding from others to sustain his brick plant simply reflects a cost– and risk–reduction strategy, prevalent among producers of artisanal brick for export. It also reflects his relative lack of specific market knowledge and contacts which translated into his dependency on more experienced and connected participants in the brick export market like Sebastián Reynoso and Jose María Cantú.

El Banco, like other rural periurban *colonias* of the *municipio*, in the post–revolutionary decades of the twentieth century evolved as a mixed tenure area with some small private properties [*pequeñas propiedades*] or small ranches [*ranchitos*] within the precincts of an *ejido*. A careful scrutiny of the lower right-hand corner of Map 2 shows this brickmaking community to be located on the southern bank of a *resaca* [oxbow lake], now a settling basin with polluted water [*estero de aguas negras*].

It is not clear when brickmaking first appeared in El Banco nor when the first kiln was built nor who built it. It is clear, however, that in the 1950s it was primarily a ranching area but that during the 1960s and 1970s it was the site of considerable brickmaking activity. Most of this activity resulted from investments by intermediaries involved in the export market or by large *ladrilleras* interested in expanding the output of their main plants [e.g., El Roble owned by

Mariano de la Fuente, *Ladrillera Mexicana* [*LAMEX*] owned by Raúl Valera, and *Ladrillera Industrializada Rio Bravo* owned by Juan José Varela].

Ascensión Navarro started working as a *ladrillero* before he married in 1948 after working over a period of many years as a *bracero* through fourteen separate contracts in California and other states, and also in Valley Fruit in Pharr, Texas. He first worked in the *ladrillera* of the Soto family in *Colonia La Copa*. He worked with Francisco, Andrés, and Guadalupe Soto for a salary of three pesos [$.60] per thousand brick molded. He also worked in the Balboa family's plant in *Ejido Los Longoria*. Navarro started working in El Banco plants as early as 1962 but after Hurricane Beulah hit in 1967 he rented land and started working on his own *ladrillera* with three of his sons. "Here we worked on rented land. We paid according to the idea of the *ejidatario*. If they did not want brick in payment we paid them in cash."

Navarro also hired workers: "Back then I worked with ten or twelve others. Those who worked with me were from Reynosa, their fathers already were brickmakers for many years. At that time this work had a better income than other jobs like construction or agriculture." In those years his clients were mostly "*Americanos*" or "*Mexico–Tejano*" truckers from Texas. He specifically mentioned Simón Hernández and Ramiro Rodríguez who he said was the last to come. He attributed the success of brickmaking in El Banco to the role of American clients. The workforce in the El Banco area in the immediate vicinity of Navarro's own brick plant also included "brickmaking people from Torreón" [*gente de Torreón, gente de ladrillo*]. He elaborated: "There were 200–300 people making brick right around here. The area that is overgrown with brush now had been cleared of brush. The workers camped there. They were from Coahuila, from Gómez Palacios. They came with their entire families." He estimated that there were eighteen to twenty kilns in his neighborhood alone but cautioned that they were moderate–sized: "The thing is that there were only kilns of five or six *bocas* [fuel ports] that held seventy to eighty thousand pieces. They are all gone now."

The market boom, according to Navarro, lasted from 1960 to 1968 and thereafter sales progressively lessened year after year. [*No duró más de 1960 a 1968. Y después comenzó menos, menos, menos, menos*]. He looked back on those years philosophically:

> Look, these things are like a game of dice. All of the big players, the ones with knowledge, that come to win, win. What does one do? Let's keep playing, right? I will play two more games and find out that the roll of the dice no longer pays off. Do you know who are the ones who keep playing? The ones who haven't won, the ones who have lost. At times they win what they have lost but it's wrong. How can you win back what you lost if there is no money? That is what happened. Times were good. The big players were winning. The slump came and only the stupid ones continued to play the game. Why do they continue to play? To waste their time.

This is a candid and realistic, if sardonic, assessment of the boom and bust market in Reynosa's handmade brick industry.

Juan González and Juan Montana, both from Pachuca, Hidalgo, arrived in Reynosa in 1962 and 1964 respectively. González found work immediately upon arrival at the Varela *ladrillera* in Cavazos where he worked for five years; and then worked for several months with Juan Tueme in Rancho Grande. He also spent some time in agricultural labor in Texas as an undocumented migrant, prior to settling in El Banco in 1970 in a house provided by Ricardo Varela's *Ladrillera El Banco* where he started working. His first child was born that same year. Montana described the situation at this *ladrillera* when he started to work, also in 1970, after working as a goatherd in El Banco for several years: "We worked real hard here from 1970 on, making a lot of brick for the other side. In the 1970s it was like a garden here, all of the land around here was for making brick. We made mostly nine inch brick. Everyone was getting involved. Varela had a plant here with his *El Banco* brand. He brought people here to work at two sites. He had four kilns. The largest had a capacity of 300,000 brick" [See Appendix 1. Photo 22 shows brick remaining from Varela's 1980s production].

When sales began to dwindle by 1981, according to Montana, "Varela stayed on simply as a buyer on credit, and I made brick for him." After his stint with Varela, Montana went to work for Pablo González—a trucker–intermediary from Rio Hondo, Texas—who had been one of Varela's clients. Juan's job was essentially that of a *campero*: he recruited and supervised brick molders, set unfired brick in the kiln, fired the brick, and then unloaded fired brick from the kiln, stacked it, and loaded it on trailers for transport. This arrangement lasted for seven years and according to Montana,

> I had eighteen people working with me for Pablo. They were all people I knew who lived around here. We had to get clay from elsewhere because it was all gone from around here. I held out as a *campero* for Pablo González for about seven years. I produced a lot of brick. He paid fifty-five dollars per thousand. He brought money on Friday to pay the workers early Saturday. When they finished working on Saturday, I paid them. I produced 25,000 brick weekly for him with all my workers. I didn't make brick myself. I set unfired brick into the kiln and fired them. That was my job. There were two kilns, one held 60,000 pieces and the other held 25,000.

From the perspective of workers like González and Montana, the particular identity of their employer was irrelevant—they could be affiliated with large plants like *Ladrillera Industrializada Rio Bravo* or *Ladrillera El Banco* of the Varelas; or they could be employed by independent truckers, some of whom like Ramiro Rodríguez from Mission, Texas had their own brand. Rodriguez's brand was *Rod's Brick*. Other truckers like Luz López and Pablo González did not have a brand and simply put the *MEXICO* mark on their brick. Ramiro Rodríguez never had a plant in El Banco but, in Montana's words, "he came in to 'help out' people. He would ask us: Do you want to make brick for me? If the answer was 'yes' then Ramiro responded: 'I am going to bring you molds.'" Montana and other small producers viewed these truckers as facilitators of their work or as those who *helped* them to work ["nos *ayudaron a trabajar*"].

José Velásquez Bauzá was an orphan raised by his aunt in Poza Rica, Veracruz. He worked at many jobs as a young boy including selling roasted ears of corn and bread as a street peddler. By the age of twelve he started working in a brick plant where clay was mechanically mixed and delivered to molders on the patio. His job was to mold and lay out [*moldear y tender*] brick on the ground to dry. He also worked as a mason's assistant [*peón de albañil*]. He came to Reynosa in 1972 at the age of twenty-two, already married and with one child. He worked for three years as a brick molder in *Ladrillera Tueme* [plant #2 in Rancho Grande] for 50 pesos [$4] per thousand and regularly earned between 250–300 pesos [$20–24] weekly but was living in rental housing.

The landlord of some of Jose's friends, Donato Gutiérrez, invited José to go to work for him at his *ladrillera* in El Banco. He paid seventy-five pesos [$6] per thousand pieces, but also provided José and his family with housing and utilities. Gutiérrez rented his plant for "ten percent of output in kind" [*pagaba 10 porciento en ladrillo*]. The plant had four kilns, one with a capacity of 48,000, and the others of smaller capacity. José was one of fifteen to twenty workers at this plant. His output varied from twelve thousand to fifteen thousand brick per month or four to five thousand weekly. In his words, "This was the most that I could make." Gutiérrez sold his brick to a client named Ezekiel, a trucker–intermediary who resold brick in Texas.

José worked with Gutiérrez for five or six years until the late 1970s when the market began to decline and the plant was shutdown. He stopped making brick and found a job with a Mexican company contracted to recondition dams and levees. That job lasted three years and afterwards José returned to brickmaking in another plant rented in El Banco by a man named Sandoval who paid him 100 pesos [$4.40] per thousand. During that period José was able to purchase rights of possession to his residence lot that included two kilns—a portion of same site that his previous employer Gutiérrez had rented.

Among his recollections of brick plants in El Banco in those years was that wives helped their husbands to mold brick. He also remembered the case of one family in which the daughters of the owner by themselves repeatedly loaded twelve thousand brick on trailers. In his opinion this work was too hard for them yet, José's wife helped him to mold brick for a time in the 1980s.

I met and interviewed Lauro Rocha in 1993 when he was working as a watchman [*velador*] at the site of Juan Tueme's long–closed plant on Praxédes Balboa street in Reynosa. As it turns out, Rocha never worked at that plant when it was operating and, in fact, had very little work experience in semi–mechanized, gas–firing plants. This reflected his deep–seated preference to work as freely as possible without strict supervision, daily quotas, or dependence upon others for doing his work. He explained his dislike of working in plants with clay mixers [*batidoras*] as follows:

> It is obligatory there, they had to make such and such a quantity for the week. They forced their bodies. A mechanical mixer produces a lot of mud and you have to force your body to work, to finish, to leave nothing because that is the task. They will pay you more if you produce so many thousands tasked for that week. You earned a little more working under the roof but you had to mold as many brick as you could from the machine–mixed batch of clay.

He understood this arrangement [*el compromiso que había*] as a trade–off: earn a little more pay as a molder in plants that had mechanical mixers and output quota systems, but lose freedom to work at your own pace with less forced drudgery. In the small plants where he preferred to work

> We made 500 daily, whereas they [plants with mechanical mixers] made 1,500 but we were free and without contracts; and after molding they had to cant and stack their brick in towers. We had to do the same but for many fewer brick, and much shorter workday. My preference was to earn less and not force my body as much as they did in the big plants.

Rocha's orientation reflected his formative work experience which was in agriculture until he was twenty years old. He was born in 1924 in the small village of Oyama in Sierra Huasteca of southern Tamaulipas but was raised, together with six siblings, in the town of Padilla near Ciudad Victoria, by his widowed mother. He had three years of schooling, learned the rudiments of reading and writing, but from the age of twelve worked as a day laborer in agriculture making fifty centavos daily. "They said to my mother, 'lend me your boy to help me plant'. So I planted so many rows from the corn that they gave me." He completed his obligatory military service and then managed to get a land allotment from the local *ejido*; he had access to irrigation and also had a ox team. However, local land squabbles forced him to leave the *ejido* and head for Matamoros in 1943–44 where he had a brother–in–law whose family had a plant where he learned to make brick. He liked the job because "it was purely contractual, you came to work at the hour you wanted and left at the hour you wanted. You simply showed up to work, and they paid you for the brick you made—100, 200, 300, 500. That was money for me, and they had to pay you, without any problem." He also worked in another Matamoros plant, one of two owned by Edmundo Lozano who had a small truck and a store where workers were granted credit against future wages.

Nevertheless, not long after his arrival in Matamoros he went across the river to work in agricultural fields in Rio Hondo, Texas to pick watermelon, radishes, beets, carrots, and other vegetables. He also worked in a brick plant in Olmito, Texas [see Map 1] but when it closed , he started working at the plant in Relámpago, Texas [ABY in Map 3]. On his trips across the river, he carried only extra clothing and, since "we were moving about as 'wetbacks' we lasted a month or two while the law was not around. You could hold on while you got some money together. I brought back twenty-five, thirty or forty dollars at a time. You always brought back a little money."

Rocha's main complaint about the Relámpago plant was that sometimes they would fail to supply molders with enough clay. Clay was shoveled onto a small dump truck from a large pit near a *resaca* but sometimes they would not supply enough for each molder who molded anywhere from 1,500 to 2,500 brick daily. According to him the dump truck had a small capacity, only enough for 1,500 brick per load, and the shovelers got too tired to supply the fifty or so molders that worked there. Sometimes he lost a day or a day–and–a–half of work due to this problem. Fortunately, since the plant was surrounded by agricultural fields, other jobs were available especially at times when crops were being harvested. The preferred job was tomato picking where experienced pickers could earn several dollars a day. Rocha was articulate in defense of the "wetback" contribution to Valley agriculture [and by extension to brickmaking]:

> We were all wetbacks [*mojados*] . . . everywhere in the fields there were pure wetbacks. And people lived as if they were homeless [*desamparados*]. That's why the harvest increased so much output in the United States. They didn't lose. Take that away and they would lose. A lot of harvests, especially of tomatoes, were lost for want of workers. The ones who are from there [Mexican Americans] worked by contract or by the hour. And the wetback who did not, the wetback over there, had to work whenever the employer said. And those who were from there, did not. As soon as the sun began to warm things up, you were there. It was help that the wetbacks were giving to the United States. There were corn harvests, the wetbacks came in to pick because those from there would not work if they were not paid so much per hour.

Rocha also complained about the output counting procedure at the Relámpago plant.

> The counter walked around and checked. Until he gave the o.k. you could not set brick. I would finish and he would come checking, checking, looking at every brick that was lopsided, that had holes in it. There were a lot of requirements. They had a long probing bar to check the wet clay batch. They stuck it in, and if it came out clean, the batch was properly mixed. If it came out muddy, they told you to give it another mixing because it was too wet. They would not allow you to mold from the clay batch until it passed the bar test. And the counter was always checking, always testing everything. Then came the guy with the probing bar.

Like many others who worked at the Relámpago plant, Rocha mentioned the foreman, Manuel Herrera, as being a religious man who ran a well–attended church at the plant site that he and other workers preferred not to attend. He also remembered that Herrera would be notified by a friend who worked for the U.S.

Immigration Service of imminent Border Patrol raids at the plant. He would warn them with statements like "Keep on your toes, boys, they are picking guys up around here. They put you on a plane to Querétaro, or to San Luís, to anywhere." Rocha and many of his colleagues would quickly walk to nearby Rio Rico and hire a rowboat to take them across the river to avoid being deported.

Following repatriation, Rocha left Matamoros for Reynosa where he had been told that one could earn higher wages than in Matamoros. "The more one moved around, the more he notices things. If they don't contract you, you are not a slave there. If you like what they are paying you at one plant you stay, if not you go looking for one where they pay more." Rocha also looked for plants where they had places to get out of the rain. Apparently, he found such a plant in the far reaches of Corrales [Map 2, lower right-hand corner] opened by Juan Tueme. In Corrales, he remembered seeing large piles of firewood everywhere and several plants, each with five, six or seven workers. There were deep clay pits on the edges of the settling basin [*estero*]. Sales were brisk, every day two or three trailers arrived to load up. There was almost no time to unload kilns, trailers were loaded directly from the kilns.

Afterwards Rocha worked in two different plants in Los Longoria, one owned by Jesus Vargas and the other by Pedro Pesinos. He stayed only two or three weeks with Vargas and spent time trying to find him to receive his wages. [*Puro batallar para cobrar lo mio, lo que ganaba*]. All of his workers were looking for him to get their wages. "Saturday arrived and passed, Sunday, Monday—and no pay." He had a somewhat better experience with Pesinos who made *adobón* for local sale in a one-kiln plant. Unfortunately, economic conditions were not good during this period so he took his wife back to Matamoros. At least, Pesinos helped him out with food and paid him on time. Rocha was comfortable making 3,000 or 3,500 brick per week but was often plagued by rain or cold weather. He adapted by trying to save money whenever he could, and by reducing expenses. For example, he would work on Saturdays and Sundays when the weather was good to make up for bad days. He summed up this strategy as follows: "If it rained on Friday and it was good on Saturday, to work. The little that I made gave me something for household expenses. And on Sunday I stayed at home. That helped my savings."

Rocha's last experience in brickmaking was with the peripatetic trucker and intermediary from Mission, Texas, Ramiro Rodríguez in *Ejido Cavazos* [Map 2, left fringe and black dot with white lettering] at a rented plant Rodriguez shared with Tano Arriaga. According to Rocha, Rodríguez did not use the plant unless he had orders from clients. "That's the way it was," Rocha explained, "he put me to work making brick only when he had an order. When he got an order, it had to be guaranteed. I think that he requested a cash advance from his clients."

Like Lauro Rocha, Luciano González's first job in the border brick Industry was in Matamoros [Map 1, lower right]. Unlike Rocha, however, González who was born in 1931 in General Terán, Nuevo León, learned to make brick working with his father and brothers in a hometown plant. In his words: "I started to work with my family in a large brick plant in Terán. They sold lots of brick in Tampico. They transported them in those years on carts. I was just a boy but I saw it. They transported brick by cart to Montemorelos and there they loaded them onto railroad cars for Tampico. It was a large plant but it closed due to declining sales." When the plant closed in Terán, Luciano, then only fifteen years old, came to Matamoros alone and found work in several wood–burning brick plants there. His parents followed him but, after his mother's death in 1945, his father returned to Terán, and González was again on his own in Matamoros. From co–workers he learned about a larger plant across the river in Relámpago, so he and some friends decided to cross the river at Rio Rico and went to work at the Relámpago plant in 1947–48. "When you work in a brick plant," González explained, "you meet a lot of people, so when we got to Relámpago people we knew were already working there."[5]

González found the plant to have better jobs and more activity than plants in Matamoros. The site was large but still had not been mechanized—clay was dug along the river, loaded by hand onto wheelbarrows, and dumped at the courts where brick were molded. Unfired and fired brick was hauled on handcarts rather than by forklifts. There were somewhere between eighty and one hundred brickmakers like himself at the plant, others employed to move brick to the kilns, and still others who set brick in the kilns under the direction of the stoker [*quemador*]. The kilns were gas–firing. He identified Manuel Herrera as the foreman, and Mr. Anderson as the absentee owner who lived in Olmito.

Payday was on Saturday and workers were paid in U.S. dollars. The molding of brick of different sizes was paid accordingly, for example, three dollars per thousand for standard size and five dollars per thousand for the large size [*barra*]. Luciano worked at various jobs in the plant over a period of three or four years. He worked about ten months each year, and usually for several days or a week at a time between crossovers to Mexico. "We went back only to visit people, our families. Then we returned again to work at the plant. You worked there all the time, if not in one job then in another. Because we could do all the jobs. I spent a lot of time in charge of shipping brick. There was a railroad line in Progreso near the plant. We loaded brick on trucks at the plant, drove to Progreso, and loaded them on rail cars." In 1952 González had to return for a while to Terán to attend his father's funeral but found upon his return that, although the plant was still firing brick, the situation was difficult for workers due to a government crackdown on illegal immigrants. He was never picked up by the Border Patrol, as were many others working at the plant. González attributed this to the fact that much of the time he was either transporting brick to the rail line in a truck, or loading brick onto rail cars in Progreso.

Inevitably, the Relámpago plant confronted a labor crisis due to the crackdown on illegal immigrant workers, and in its last years of operation imported unfired brick from fields across the river from Progreso to fire in kilns at the plant and, finally, bought brick molded in fields in Corrales or El Banco, had them transported to kilns rented from Jesus Vargas in *Ejido Rancho Grande* for firing, and then loaded them onto plant trucks and hauled them back to the plant in Relámpago. During this final period, González was in charge of firing the rented kilns and transporting fired brick for Manuel Herrera who ran the business until 1956 when it shutdown (Cook 1998:94–97).

In 1957, González went to work for Gustavo García in Mexico at a plant he had near the junction of the natural gas line and the highway to Ciudad Rio Bravo before he became a partner in *Ladrillera Reynosa*. He remembered that Ramiro Rodríguez, long before Rocha worked for him in Cavazos, bought fired brick at Garcia's plant and tried to hire him to help operate a plant that Rodríguez was planning to open, and subsequently did open, in Hidalgo, Texas. González declined Rodriguez's offer and decided to remain at Garcia's plant where he became, as he explained, a sort of managing partner: "Gustavo said to me, 'Stay here and manage it,' and he got me a bank loan and I went to work. According to him we were now partners. Everything went well because I managed it well. After some time had passed, Gustavo told me, 'Look, you are the industrial partner, I am the capitalist partner, you manage the industry, I manage the capital'." By 1959–60, however, González found that he was working too much without seeing any profits, and he decided to pull out of his arrangement with Garcia Garza: "I could not work with him any longer because practically I was not really a partner, I was a worker because I did not have anything at my disposal [*No podía disponer de nada*]."

After ending his relationship with García, González went to Camargo to work in the Salvatore Grimaldi plant in 1960 and then for Ramón Longoria until 1967 when Hurricane Beulah struck. He had purchased some land with two houses in Camargo and, in the storm's aftermath, he agreed to sell out to Longoria, whose plant already surrounded his land and who he did not like as an employer, for 25,000 pesos [$2,000]. Next González went to work for Magín Villaseñor in his Cavazos plant for about a year and a half until it was sold to Varela, then worked for Raul Valera in Rio Bravo until around 1972. Until his retirement from the brick industry in 1984 he operated a plant, previously rented by Magín Villaseñor, belonging to Salina de Gracia who also owned a local chain of fried chicken restaurants. Among González's clients at that plant was none other than Ramiro Rodríguez, who had become his friend over the years, and who he accompanied on delivery trips to Houston and other Texas cities.

Brickfield Bosses, Subcontracting, and Rent

During the boom days of artisanal brick production for export to Texas in the 1960s, 1970s and early 1980s, a large volume of handmade brick was produced through the *brickfield boss* [*campero*] system—an arrangement in which a knowledgeable brickmaker was subcontracted by his employer, the owner of a large plant, to recruit molders and to supervise them in molding unfired brick on rented land outside the confines of the main plant. The unfired brick made in the satellite field sites were transported to the main plant to be fired. In some cases wood–firing kilns were also built at satellite field sites.

It was also common, however, for large, gas–firing plants to subcontract small, wood–firing plants to produce fired and/or unfired brick for them—either with the big plant's brand name or with a generic *MEXICO* identifying mark. This practice bridged the major division in the industry and, in fact, fostered a dependency of the small plants upon the large plants. Owners of these small plants involved in such a sub–contracting relationship were also referred to as *camperos*.

Since all of the large, gas–firing plants in the greater Reynosa riverine zone operated satellite *campos* during a period when union organizing efforts were focused on their main plants, the question is also raised as to whether this overlap was coincidental or was reflective of a conscious strategy by the big plant owners to use sub–contracting as a means to decentralize and disperse production and, in the process, to complicate unionization efforts.

The first examples of satellite brickfields in the modern history of the border brick industry involved Texas–based *ladrilleras* with kilns in Texas but brickmaking operations in Reynosa where land was leased or bought for that purpose. This arrangement was described above in the case of Anderson's Relámpago plant but also occurred in other cases, perhaps most prominently involving Guenther Weiske's *Valley Brick and Tile Company* in Madero, Texas that throughout the 1950s and into the 1960s imported unfired brick from Reynosa for firing in its own gas–fired kilns (see Cook 1998:110). The cases of Anderson's Olmito and Relámpago plants [*Aztec* brand] and Weiske's Madero plant are the clearest examples of *maquila*, twin plant relationships, precursors of the *maquiladora* arrangement institutionalized through the Border Industrialization Program in 1965 for electronics, clothing, and other industries.

There are other cases during the 1950s as well. One of the earliest was already described in chapters 1 and 2, namely, the plant in Hidalgo, Texas involving several associates: Trinidad Herrera, Lester Tisdale, Ramiro Rodriguez, Bill Cypert and Magín Villaseñor. According to Villaseñor, who had previously hauled fired Saltillo brick to Hidalgo for Tisdale, Tisdale began to fire brick [in Hidalgo]. Villaseñor was one of the first to transport unfired brick to Hidalgo to be fired. He did not yet have a kiln at his Reynosa plant, so he arranged to fire them at Tisdale's plant in Hidalgo. In short, Villaseñor replaced Tisdale's Saltillo supplier and became a brickmaker himself instead of a hauler. Since the land Villaseñor purchased on the outskirts of Reynosa did not have a well, he was forced to buy water from tank truck service in order to prepare clay for brickmaking. As he explained:

> There was no well on the property. It was necessary to cart water from a tank truck to a cement reservoir. Not even the clay was dug on the property. It was

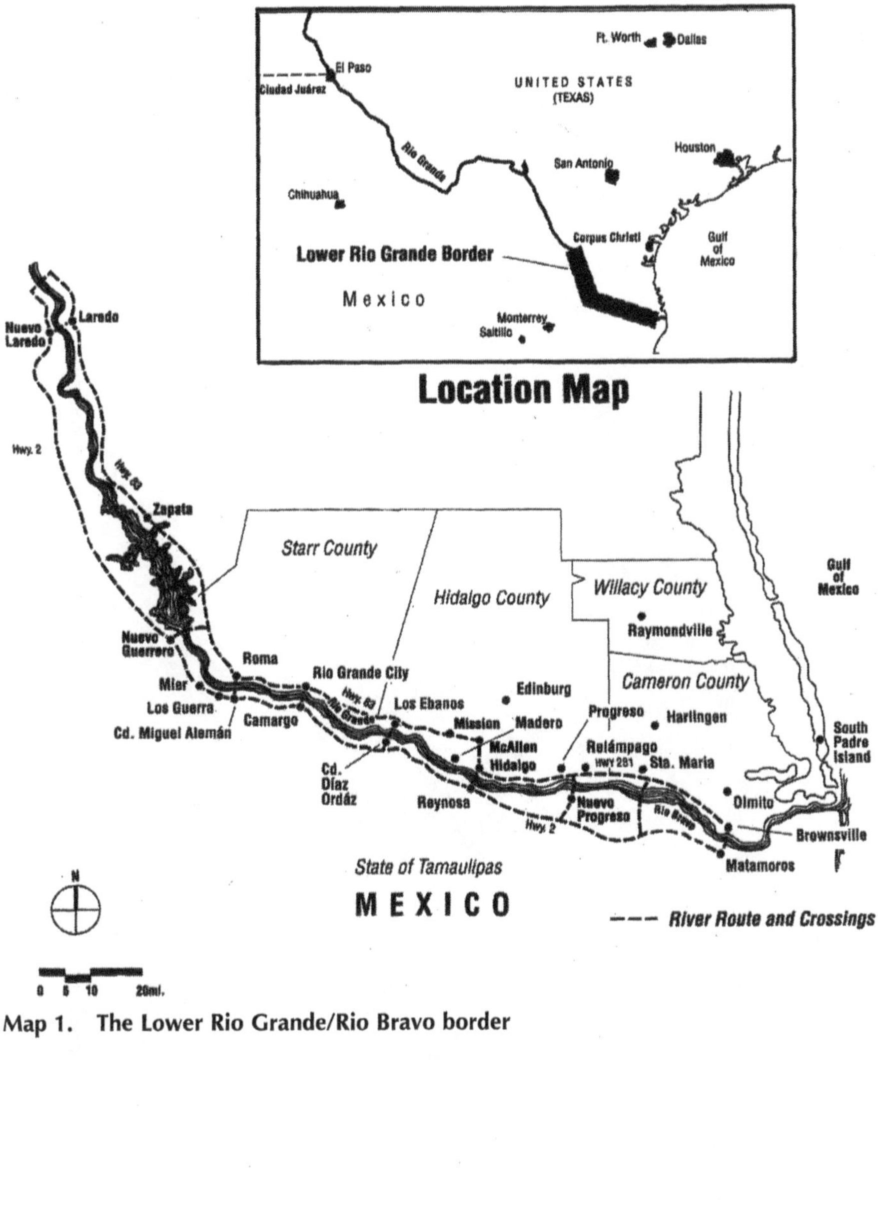

Map 1. The Lower Rio Grande/Rio Bravo border

Map 2. The riverine brickmaking belt of the municipality of Reynosa and Hidalgo county

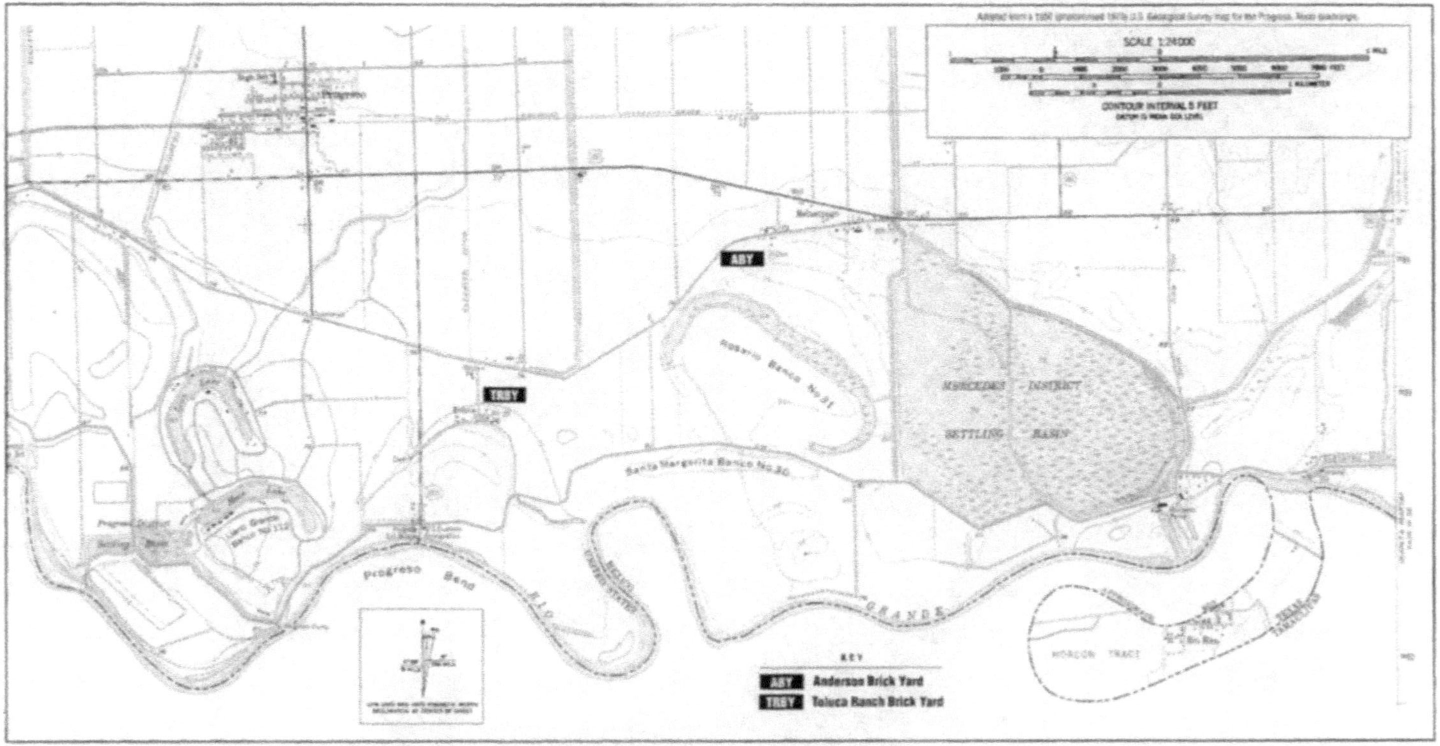

Map 3. The riverine/deltaic topography of the Relampago-Progreso area of Hidalgo county

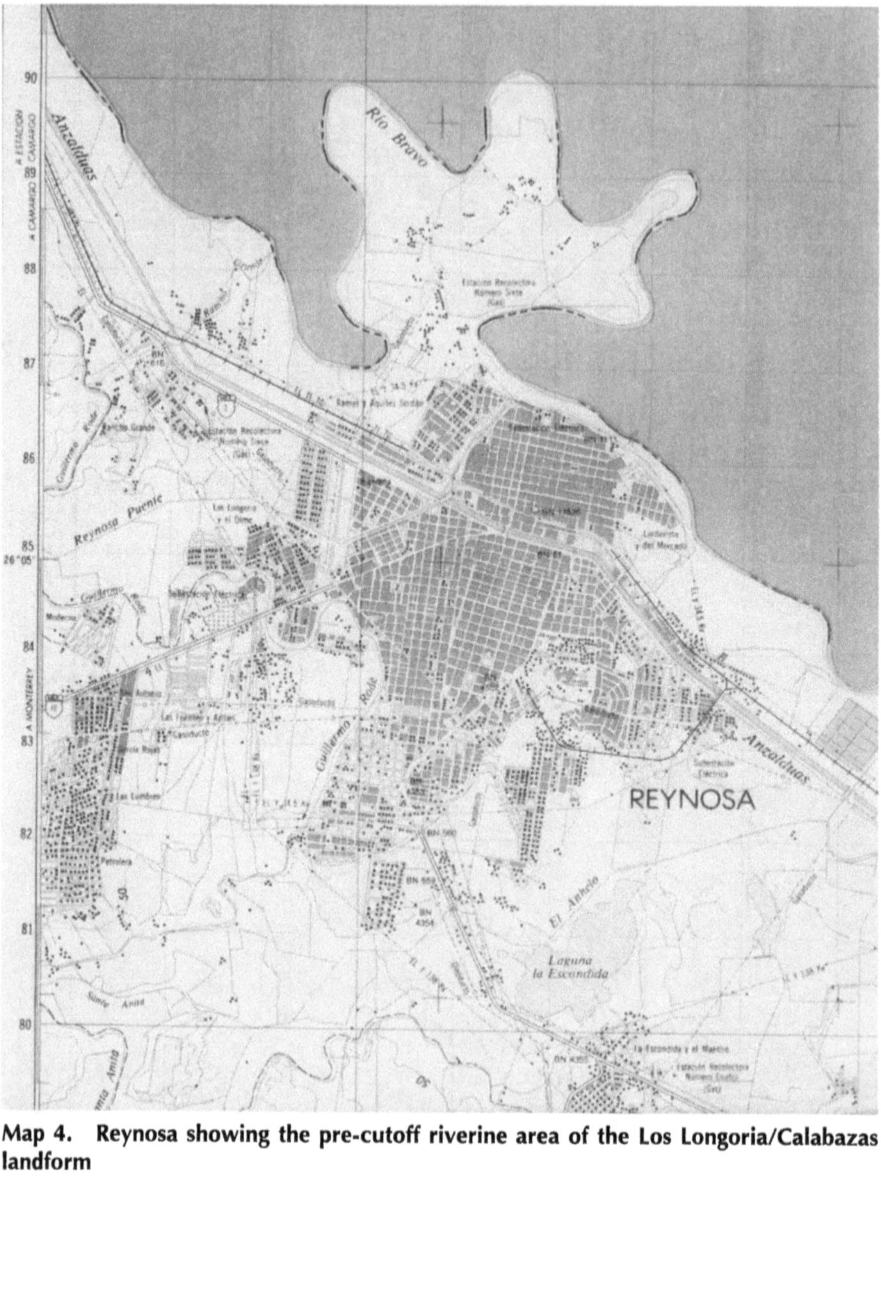

Map 4. Reynosa showing the pre-cutoff riverine area of the Los Longoria/Calabazas landform

Map 5. The lower Rio Grande/Rio Bravo showing the Pate and Rincon bends and the pre- and post-cutoff boundaries of Reynosa's Los Longoria-Calabazas landform

> transported in a dump truck from the riverine zone. I went to Saltillo and brought thirty men to make brick for me. I rented them houses and I started to make brick. Villaseñor remembers storing his first unfired brick along the front portion [*cabecera*] of his property that still had corn stalks and milk cows [*había zacatito, había vacas con leche*].

It wasn't long before Villaseñor dug a well at his first plant to avoid having to buy water for mixing clay. As he acquired more vacant lots on Reynosa's outskirts to expand the output of unfired brick for Tisdale's Hidalgo plant, he earned the nickname '*Las Norias*' because the first thing he would do on newly acquired parcels was have wells dug.

Tisdale, at some point, decided that it would be more advantageous to import fired rather than unfired brick. So, he persuaded Villaseñor to build a kiln on the land rented for the patio. This was Villaseñor's first kiln and he fueled it with firewood. But Tisdale was getting increasingly involved in the cement business, importing fifty trailer loads of cement daily from Monterrey, Nuevo León into Texas through Reynosa, Miguel Alemán, and Nuevo Laredo. As explained in chapter 1, Tisdale was able to monopolize the output of Monterrey's cement industry, causing a shortage of cement in Mexico and, ultimately, an embargo on the export of construction materials, due to formal complaints filed by the Mexican construction industry with the federal government's *Secretaría de Industria y Comercio* [Department of Industry and Commerce].

When the building materials embargo went into effect at the Reynosa–Hidalgo bridge Villaseñor had just finished unloading his first load of brick fired in his newly constructed kiln. He improvised a solution by borrowing the truck of Tisdale's partner, Ramiro Rodríguez, loaded it with brick at his Reynosa plant and then drove down the road to Rio Bravo so that he could cross the bridge at Nuevo Progreso into Texas where Rodriguez would meet him. Villaseñor had speculated that Mexican Customs was enforcing the building materials embargo only at the Reynosa–Hidalgo bridge, and would not yet be doing so at the Rio Bravo–Nuevo Progreso bridge. His hunch was correct, and he was able to temporarily transport his brick into the U.S. despite the embargo. It turns out that the trade embargo was never intended to apply to brick, only to cement, so it was officially lifted within a week, much to the relief of the *ladrilleros*.

In summary, what began as a *maquila* arrangement to make crude brick in Tamaulipas fields for firing in Texas kilns, evolved into a Reynosa–based brickmaking enterprise, somewhat paralleling the case of Sebastian Reynoso with *Aztec* brick and Anderson's Relámpago plant. Like Reynoso, Villaseñor's enterprise grew rapidly to include multiple plants.

A significant proportion of the piecerate molders who worked in the campero system were temporary or periodic workers, often recruited and transported either by the *campero* or by labor recruiters [*coyotes*] from off–border areas of handmade brick production—especially in the Torreón area of Coahuila and the area around Monterrey. It was typical for members of this *campo* labor force to use employment in the Reynosa *ladrilleras* as a stepping stone for a "jump to the other side" [*brinco al otro lado*], that is, to enter the undocumented labor force in Texas.

Considering the situation of the artisanal brick industry in Reynosa from the 1950s on as it became increasingly dependent on the masonry construction business cycle in Texas, when the demand for brick expanded either existing independent household producers found ways to expand output, or there was simply an increase in the number of such units. The net result, either or both ways, was an increase in handmade brick output.

Also, the gas–firing plants were able to selectively mechanize clay preparation so as to give already employed brick molders more time to mold, or they hired more molders if the existing plant had underused capacity. If at capacity, output of a capitalist brickmaking enterprise could still be expanded through the *campero* system in which a *maestro ladrillero* was contracted to produce unfired brick to supplement current production of an already existing firm, typically on rented or leased land. The *campero,* who was also a *mayordomo,* was responsible for recruiting workers, supplying them with access to the necessary raw materials and tools, and for organizing and supervising the production and transport of unfired brick to the kilns of the main *ladrillera.* This system drew on a nearby periurban or more distant rural casual labor pools, typically in areas like Monterrey and Torreón where there is known to be considerable handmade brick production. Either the *campero* himself or a labor recruiter hired by the *campero's* employer would travel to the off–border area to recruit workers, offering cash advances, transportation, food and housing, and a higher piecerate to provide incentives to the off–border workers.

The Tex–Mex Plant and the Field Boss System

The *Tex–Mex de Mexico, S.A.* plant, located on the riverine highway in Cavazos just below *Ladrillera Industrializada Rio Bravo* and the PEMEX petrochemical plant, operated for a decade starting in the mid–1970s, first as totally dependent upon off–site field production but later also began on–site production [see Map 1—the second of the three dots on the lower left side shows the approximate location of this plant]. Two informants, Omar Zapata and Antonio Méndez who were interviewed separately, were employees at this plant from its 1974 opening until its closing in 1985, Zapata as plant and sales manager, Méndez as foreman and kiln stoker. Prior to the *Tex–Mex* acquisition of the eighteen–hectare site, it had already been developed as a brick plant with a natural gas line and several kilns. According to Méndez when he and Zapata first visited the plant site it was being run by Raymundo García who leased it from Juan José Varela, the original owner. David Brachot of Brachot Lumber Company in Houston acquired the plant from Varela. It was not in good shape; the four existing kilns of 150,000 piece capacity each required renovation or rebuilding, and molding/drying sheds and other roofed structures were either renovated or rebuilt. Méndez said that they managed to repair one kiln in short order and to fire it within a month after the take over. Subsequently, two additional kilns were built ranging in capacity from 200,000 to 250,000 pieces. This meant that the renovated plant had a total of six kilns with an approximate combined capacity of one million pieces. During peak months, according to Zapata, the plant produced 2,800,000 brick, about one half of which came from off–site fields.

Zapata depended entirely upon Méndez to staff the plant since he knew that Méndez, prior to working for many years at *Ladrillera Reynosa* where they met, had worked in several other plants as well and, therefore, knew many brickmakers. Antonio Méndez was born in Dolores Hidalgo, Guanajuato in 1935 and came to Miguel Alemán in 1950 with his father and four brothers "to look for a better life because here on the border you earned more cents" [*buscar la vida, porque aquí en la frontera se ganaban más centavos*]. He never attended school and, at the age of twelve, was working on railroad track maintenance in Guanajuato.

Méndez was proud of his ability help operate a large brick plant, and count inventory for millions of brick, without formal schooling. He candidly admitted, however, that counting inventory was not easy for him:

> Every month I did an inventory count. I did a lot of counting, a lot of counting, and I didn't even have a culculator. It was just from my head. There were times when I spent two hours counting and I got stuck, and didn't know if I was going backwards or forwards. I made a count and saw that it did not come out right so I would do the count again. Sometimes I would come and lay down for twenty minutes or so, or I would drink a soda to cool down my head, and afterwards I was able to make a count. I counted more than a million brick at one time, and they were not all together in one stack but in many separate stacks.

The Mendezes had not worked in brickmaking in Guanajuato but during their first year on the border combined ranch work with work at various brick plants. Until 1953 when he finally came to Reynosa, Antonio worked in several plants in the Miguel Alemán area, and even spent several months working at the *El Gallo* plant near Monterrey. He preferred to vary his job at these plants to avoid boredom—working seasonally as a molder and then as a setter or drawer, and so on. In Reynosa, he worked for Alberto Gómez, Magín Villaseñor, Juan José Varela, Efraín Salinas, and several months in Ciudad Rio Bravo prior to going to work for Joél Rodríguez at *Ladrillera Reynosa* in 1970. He learned how to fire kilns at *Ladrillera Industrializada Rio Bravo* over a two year period starting in 1968, so was able to work as assistant or night–time stoker at *Ladrillera Reynosa*.

Méndez, in elaborating on production at the plant, mentioned an important point that Zapata overlooked, namely, it had a gas hookup suited to firing only one kiln at a time at full heat, and another at a low heat for drying purposes only. This meant that only one kiln weekly could be fully fired. The production schedule was to try to keep several kilns all the time so that when one was taken off line another was immediately put on line and fired. As Méndez described the schedule:

> We had five kilns. One or two were firing, others were being filled. One or two were cooling down. Others were resting. We had two teams setting brick, another team of six men were drawing brick, the rest were still hot and were cooling down. It took a team of six about one week to set 200,000 brick. Cooling down takes an average of twelve days. Firing takes about the same—ten or eleven days according to the condition of the brick in the kiln. If brick are set that are not uniformly dried, it takes longer for them to fire properly. That impacts on the average number of days to complete the firing process.

He also admitted that the production schedule was disrupted at times when orders backed up. This caused them to put two teams into drawing and two into setting.

The gas line problem put Tex–Mex at a disadvantage with *Ladrillera Reynosa* where, according to Méndez, four kilns could be put on line and fired simultaneously. At Tex–Mex molding and drying production was significantly larger than firing capacity. This meant that the peak month production estimate given above by Zapata of two million eight hundred thousand, if accurate, had to include both unfired, and fired brick.

Omar Zapata's two–decade career as a sales and production manager in the Reynosa brick industry spanned the boom years from 1966 until 1985, and put him into the thick of the hectic surge in industrial activities involved in supplying a galloping demand for Mexican handmade brick in the Texas building market. Before working at Tex–Mex, Zapata was with *Ladrillera Reynosa* where he learned the complex details of production and marketing that he used to his advantage at Tex–Mex. When I interviewed him in 1992, seven years after his retirement, Zapata talked enthusiastically and articulately about what he repeatedly referred to as the "fabulous" brick business in the years of high demand because "you could earn what you wanted to earn in just a few years." The period he is referring to started around 1968 when during meetings of the Association calculations were made that Reynosa's plants alone—not taking into account those in Rio Bravo, Matamoros, Camargo, and so on—were exporting thirty-five to forty trailer loads [455,000 to 520,000 pieces] of brick daily. "Everyone began to understand that there was a boom, that it was a fabulous business." Zapata was proud of the quality of the brick he produced, citing his attractive twenty year old ranch style house in the center of *Colonia Rancho Grande* [where the interview was conducted] as an example of the durability and aesthetic appeal of Mexican handmade brick.

The Tex–Mex plant, like *Ladrillera Reynosa* and the other large gas–firing plants, trucked in clay from land near the river. Pits ranging from one and one-half to three meters deep were dug mechanically and enough clay was extracted to keep the molding process supplied. According to Antonio Méndez, this meant that 180 truckloads of clay were delivered to the site weekly since each molder molded an average of three to four truckloads weekly and there were usually sixty molders working. There were four or five trucks running all day and making eight, nine or ten round trips daily between the riverine clay pit and the plant site. Upon arrival at the plant, the clay was processed in a mechanical mill to remove gravel and to produce a consistent texture that reduces fracturing of brick during the molding process. Zapata pointed out that unlike other areas of interior Mexico with softer, purer clay soils, the riverine clay of Reynosa has lots of sediment and sand like particles that require more heat to harden. Once the clay was milled it was dumped at the edge of the roofed sheds where each molder mixed the clay with water to prepare his clay supply for molding. Essentially half of the mound prepared on the first molding day of the week [usually Tuesday] is used to mold brick that same day, and half is kept wet overnight and used the next day, and so on successively until the last batch for the week is molded [usually on Saturday] on two successive days.

Unlike *Ladrillera Reynosa*, Tex–Mex did not have a mechanical clay mixer so the sifted and milled clay was brought to staging areas near the roofed molding and drying sheds where each molder every afternoon would prepare his own clay batch with water, let it sit overnight, mold an average of 800 brick the next morning, lay them out to dry on the shed's drying floor, and then prepare a clay batch for the next day. Méndez supervised anywhere from fifty to seventy molders. Daily output varied from 600 to 700, or in some cases less, to as many as 1,000 in some cases. Daily output of more than 1,000 daily, especially when the molders have to prepare the clay, was abnormal. A weekly output of four thousand pieces from a molder was, from his perspective, ideal.

During the drying process, the brick are turned over on their sides, then are stacked in towers crosswise and lengthwise in alternate layers with space between them to enhance the circulation of air and to promote drying. Two advantages to conducting the molding and drying processes in roofed sheds are limiting the damage from rainfall, and from the summer sun which causes drying brick to crack.

Dried, unfired brick were trucked into the kilns where they were hand set in a special way to maximize the flow of heat and ventilation, and then fired carefully so that the brick heat slowly until the fire moves to the top of the stack which is covered with a layer of fired brick.

Zapata admitted that the Tex–Mex plant depended upon the field system for about fifty percent of its output, every day, every week, every month, every year. He managed ten to fifteen small fields. Most of the time unfired brick from the fields arrived in small truckloads of two thousand pieces. Each load had to be examined by Zapata or Mendez to make sure that the brick were properly molded. "What we called *campos* [*fields*]," Zapata explained, "were people that supplied us unfired brick from along the river. We paid them so much per piece or so much per thousand pieces delivered to our plant. We set them into the kiln. The *campero* hired people, paid them a determined wage, paid the transport to the plant with our brick, and kept the rest for himself."

Interestingly enough Zapata, while plant manager at Tex–Mex, started his own plant on his father–in–law's land, and built four kilns, three of 60,000 capacity each, and one of 100,000 capacity. He was doing a brisk business in selling fired brick in direct competition with Tex–Mex. He had twenty to twenty-five employees. Subsequently, he made a deal with Tex–Mex to sell them all of his output in unfired condition. At that point he himself, in effect, became a *campero*.

From Zapata's perspective, one of the main advantages of the satellite field system was that it provided a way to increase output without increasing the number of workers and attendant labor problems at the main brick plant itself, while simultaneously transferring production risks to the *camperos*. As he expressed it:

> The field bosses got their field. They employed people to make brick. After the brick was made the field boss had to pay the workers. If the brick got wet it was the field boss who took the loss. But if it didn't rain, the field boss hired a truck, and loaded the brick. He earned twice what it cost him to produce them. He set the brick in the kiln of the plant where they were fired. That system helped us avoid many problems because if we had another forty, fifty, or sixty workers at the plant we would have had to pay social security. The *campero* system was good business.

Precisely because of these and other advantages, the *campero* system flourished during the 1960s and 1970s but it was highly competitive, pitting Association members against each other and against the horde of intermediaries. Zapata described this competitiveness over *camperos* as follows:

> Everyone had fields. More money and they took your field away from you. I had Tex–Mex but the foreman of Reynosa Brick would go and take it away from me, or the foreman of Salinas, or Tueme, or Morelos, or Longoria—all of them had fields. There was lots of action. We had a vehicle assigned to drive all day around the fields to make sure that they were still working for us.

Zapata went on to clarify that it was not so much the output itself that the *campero* sold to competing enterprises but the capacity to produce a given output. He explained that a field boss could not sell brick already made with the brand name of a particular plant but once the brick were molded and delivered he could return one plant's molds and then get molds from another plant.

In summary, the *campero* system, was created by a combination of trucker and other intermediaries and by the largest brickyards to augment their supply of brick during the boom years of the 1970s and early 1980s. This system explains, for example, the wide geographical dispersion of particular brand names in brickyards throughout the riverine periurban area of Reynosa. Wooden molds, complete with attached metal templates [*placas*] with a particular brand name which is impressed into the face of each brick enclosed within an indented rectangle or frog one–fourth to one–half inch deep, were supplied by the sponsoring *ladrillera* to the *campero* who, in turn, distributed them to the molders working under his or her supervision (see Cook 1998:20–21). Also, small plants were sub–contracted by trucker–intermediaries, as well as by large plants, to supply either export–ready fired brick with the generic MEXICO mark for Texas based intermediaries like Luz López, or unfired brick stamped with the brand name of the subcontractor, for example, *Ladrillera Reynosa, Ladrillera Morelos, Ladrillera Industrializada Rio Bravo.*

Trucking and Wheeling–Dealing During the Boom

Although the arrival of rail freight service in the Rio Grande Valley at the beginning of the twentieth century did provide a means for brick produced on the Mexican side of the border to be shipped to Corpus, Houston, San Antonio, and beyond, it was trucking that was critical to the take–off of the Mexican brick export industry. Indeed, after the ending of rail freight service by the mid–1950s, there are several examples of key players in the industry who were brick haulers prior to becoming involved in production, for example, Magín Villaseñor and Elizardo González; or who got involved in trucking subsequent to being producers like José María Cantú. In these and other cases such as Raul Valera, and Juan José Varela, involvement in general freight hauling was combined with brick hauling as well as with brick production.

From the mid–1950s, then, trucking has been a focus of the competitive struggle between Texas and Mexican brick producers, as well as a periodic source of crises or problems for all parties involved in the export–import of Mexican handmade brick. It is precisely in the cross–border traffic of brick hauling trucks that is most vulnerable to the *de jure* and *de facto* contradictions and complexities of U.S.–Mexico commerce. At play here are many factors: the respective laws of the two countries regarding foreign and domestic trucking which involve everything from the registry and credentials of the trucks and their drivers to the conditions or restrictions on truck movement once they have entered the other country; the dynamics of enforcement by customs officers at specific ports of entry so that what is 'par for the course' at one port of entry may not be so at another; and the lobbying efforts of private or special interest groups like the Texas brick producers or the Mexican trucker's union which can impact upon how or when particular laws or regulations are or are not strictly enforced vis–à–vis brick hauling.

Many of these issues, from the perspective of the Mexican brick producers, are expressed in a 1981 letter from the Reynosa–based *Asociación* to the Director General of the *Insituto Mexicano de Comercio Exterior* [IMCE], the Mexican government's principal agency for setting and enforcing export–import policies and regulations. Below is the full text of this letter [author's translation]:

> We are turning to you to inform you about the situation that for several years has been confronting the brick producers of Reynosa who export our brick to the United States; to begin we must tell you that the majority of us do not have trucks to export our products especially to the places where they are consumed. Those of us who are dedicated to sell in the export market either have or contract vehicles from the United States and due to our geographic situation these trucks must enter 10 or 12 kilometers into Mexican territory [to haul brick to Houston, Corpus Christi, San Antonio, Dallas, Fort Worth, etc.].
>
> Since our laws do not permit American trucks to come to buy Mexican brick in the areas mentioned, that is, the Office of Public Finance and Credit or, specifically, Customs or the Federal Vehicle Registry, interpret and enforce this law.
>
> Moreover, the federal authority does not permit drivers, either Mexican or American, to drive without carrying their respective passports. Lastly, the Office of Communications and Transportation, does not permit American trucks to transport these products in Mexico. For that reason we are at the mercy of these government offices to permit us to transport our products legally to the United States without acting in accordance with the Law, cooperating with us for the purpose of obtaining foreign currency, a situation that we are in disagreement with because at any time that functionary who wants to enforce the letter of the law will detain trucks, provoking chaos in the United States so that other clients will not buy brick from us. Already on several occasions detentions have occurred with the result that little by little they stop coming to buy brick in Mexico.
>
> To have the possibility exist of not having to ask favors of all the above–mentioned government offices, we are requesting these authorities to modify the respective laws, or simply to issue a memorandum to the appropriate customs officials. In the absence of this we will continue approaching said customs officers to request that they permit the passage of these vehicles so that they can come to transport brick, an incorrect procedure that puts us at the mercy of particular customs agents.
>
> To conclude, we do not have the means to buy the vehicles in question. First, it takes a year to have them delivered. Second we would have to sell our brick plants in order to buy the vehicles. Third, it is not cost effective to do so since it would take so long to pay off the vehicles in order to get a return on the investment on them.

The implicit theme in the third and fourth paragraphs of this document is the informal cost of getting permission for brick–hauling trucks to move across the international bridges. The phrases "asking for favors" [*pedir favores*] or "to be pleading" [*estar suplicando*] customs employees [*empleados aduanales*], activities considered to be "unseemly" [*muy incorrecta*], are thinly veiled or euphemistic references to the need to pay bribes [*mordidas*] to Mexican Customs officials in order to keep the brick traffic moving. As Ramón Longoria, owner of three *ladrilleras* in Camargo, *Ladrillera Riberena, S.A.*, *Ladrillera Mecanizada Riberena, S.A.*, and *Ladrillera El Alamo de Camargo, S.A.*, which eventually were consolidated into one enterprise, *Ladrillerías Camargo, S.A.*, candidly admitted: "I paid bribes . . . brick were officially free of duties but without the bribe there was no *go ahead.* Then one would make demands and you would make arrangements with another. I didn't want to report them but that's the truth. It was a shame." Because of this situation, Longoria, during meetings of the Association to develop a counter–strategy to the BIT [Brick Institute of Texas] campaign, argued against lobbying the Mexican federal government to put inspectors on the bridges who would keep the unfired or under fired brick from entering Texas. He argued to his colleagues that this would result in "other bribe takers to extract bribes from you claiming that the brick is bad." He summed up his position to his colleagues as follows: "Every time you ask the government to intervene, you are going to have to gain the favor of another bribe taker." He added that all of the members of his producers' association said that "they conducted business in spite of the government." Longoria was not a signatory of the 1981 letter. Unlike them, his enterprise had a fleet of twelve trailers to haul brick from the Camargo plant primarily to Houston.

Over the decades there were a variety of trucking arrangements in the border brick industry. On balance more brick were hauled by independent truckers than by the producers themselves with their own trucks. And even in the latter cases, independent truckers also did business with the these producers. Regarding truck registry and the nationality of the drivers, the situation also varied. A typical situation was that of Elizardo González, a Mexican national who started out in the brick industry as a trucker but eventually became a producer; his six trailers were registered in Texas but were driven by Mexican nationals. He hauled loads of brick from his and other brick plants in Reynosa to Houston. Then, on the way back from Houston he would load up with chicken feed which could not be unloaded in Texas. He explained: "I put González Trucking on my trucks. I could load up in Mexico and cross into Texas loaded. I could load in Texas but only to leave Texas loaded. I could not unload in Texas. I had to load for Mexico and then load for the United States."

Trucks of Mexican registry with Mexican drivers also participated in brick hauling—so long as they could deliver their loads and return to Mexico within a twenty–four hour period after paying a $12.50 fee to U.S. Customs. This seems to have been a widespread practice in the 1960s and 1970s, and presumably was advantageous to the Mexican producers and to the U.S. importers since it enabled them to use cheaper Mexican diesel for at least a portion of their trip and also exempted them from having to meet more costly and more stringent state and federal trucking standards and regulations.

Ed Stout, President of Acme Brick, spoke candidly about the independent trucker as both a scourge of the small Mexican producer as well as a source of unfair competition for companies like his in the Texas market (interview with author 11/5/92):

> Another thing that hurt Mexican brick, people work their butts off making those brick, you talk about hard work, little bitty kids, hard work, and they would be at the mercy of truckers. The independent Mexican trucker would come in and, if the situation were right, would get the brick for practically nothing. Then they would drive with nine cents per gallon fuel. They never heard of DOT's [Department of Transportation] sixty hours a week maximum driving time or a driver's daily log filed with DOT because *No habla ingles*. And they'd get to Houston and they'd drive up and down the street like they were selling watermelons. They are successful when there's a shortage of brick.

The predominant pattern, however, was for an independent Mexican–American trucker or a Spanish–speaking Anglo–Texan trucker, with Texas–registered trucks, to pick–up brick in Mexico for delivery in Texas. Many of these truckers were hired by or worked for brick dealers, whereas others were independent truckers and brick dealers.

For several years from the late–1960s to the mid–1970s, the dominant intermediary–trucker in the border brick industry was Chuck Revelette who operated a fleet of more than fifty tractor–trailers that plied the border from Matamoros, Tamaulipas to Piedras Negras, Coahuila. Although his involvement in the border brick business was relatively short, his impact was exceptional due to a combination of factors, not the least of which was his larger than average size, colorful and forceful personality, hard–driving competitive business style, and an aggressive business plan to corner the Mexican handmade brick market. He was, by all accounts, the preeminent wheeler–dealer in an arena where they proliferated.

Born in Pawhuska, Oklahoma in 1936 into an Osage tribal ranching family, Revelette entered Texas A & M in 1954, left after his first year for a two–year tour of duty in the Marine Corps, then returned to play defensive tackle on Bear Bryant's football team, and in 1961 graduated first in his class with a degree in animal husbandry. Fresh out of college as, to use his own words, an "educated cowboy," he was hired by Octaviano Longoria to manage a 900,000 acre ranch and 17,000 head of cattle in the states of Tamaulipas and Nuevo León. He enjoyed this work because it involved doing "everything mostly on horseback, like it was in the old days, including a cattle drive originating in Los Múzquiz, Coahuila." He recalled making an $800,000 profit on cattle he bought on a trip to the state of Veracruz and shipped by rail back to Longoria's ranch for fattening and resale. Reluctantly, Revelette had to quit this job after five years in the aftermath of Octaviano's death to avoid having to take sides in what he called a "pissing contest" among the four surviving Longoria brothers.

By the time he left the Longoria ranch in the mid–1960s, Revelette had achieved relative fluency in Spanish, which as it turned out, would be a major asset to him from that point on in his career. During his five years in Mexico along the U.S. border he was impressed by the considerable activity he had observed involving the transport of handmade brick to Houston from several Mexican plants including the 'O.K.' brick plant operated by an American expatriate named Jack Temple in Pesquería near Monterrey. So, with his ranching career ended, Revelette traveled to Houston to look for a job in the brick business. He was hired as a brick salesman by the Great Southern Supply Company, a Houston distributor for the Henderson Clay Products Company that, at the time, did not import Mexican handmade brick. This is a situation that Revelette remedied once he gained familiarity with the Houston market and convinced Great Southern's owner, Mr. Bashears, that there was money to be made in the Mexican brick import business. After Hurricane Beulah devastated many brick plants, Revelette was in Reynosa to establish new business contacts but ended up making them in the local 'boystown' [*zona de tolerancia* or red-light district] where he and a business associate waded from brothel to brothel to make the best of a bad situation. Until the border brick industry recovered from the hurricane damage, Revelette imported brick from the Monterrey area.

By the late 1960s Revelette's success with Great Southern, achieved mostly through his importation of many carloads of brick by rail from Mexico, provided him with the financial means to start–up his own business in Houston that he named Metropolitan Brick Company. Initially, he attempted to have brick delivered by rail from Mexico as he had done at Great Southern. But this proved to be unsuccessful due to excessive demurrage costs, forcing him to turn to truck transport. The business began modestly but grew rapidly. Raúl Valera, who remembered Revelette as being an intelligent and hardworking young man, first did business with him at Great Southern but later sold Metropolitan 200,000 brick monthly from his Rio Bravo plant.

According to Lester Tisdale who was a salesman for Revelette at Metropolitan in the early 1970s, "He [Revelette] had it set up to where we had two boom trucks, a forty–foot trailer, and a short bobtail. The trucks would come in at night, and he would unload them at the yard or keep them on the truck for delivery the next day." Thanks to hard–driving, 'never–say–die' sales efforts, long hours, low prices, good quality, and reliable on–time delivery to job sites, Metropolitan built–up a client base that included many of Houston's largest builders, like Suburban Homes where he supplied brick for twelve hundred to eighteen hundred houses annually. In the latter case, Tisdale described how he got no orders after five or six Thursday visits with the firm's buyer, but on the seventh visit was rewarded with a challenging 'do–or–die' opportunity to supply brick for ten houses at the job sites on the same street in a new subdivision by the following Monday morning. The requested brick was delivered on Sunday and, as a consequence, for several years Metropolitan supplied about one million brick annually to Suburban Homes.

Revelette's business plan was bold and fast–tracked. He filled his initial orders from a plant in Rio Pesquerías, near Monterrey, where he had convinced the owners to consign their entire output to him on credit in return for a share in profits from his sales. In his words: "I bought everything they had on credit. They financed me. That's how I got started." As his client–base and earnings grew, and his network of suppliers expanded, Revelette went to White Motor Company and purchased on credit a fleet of fifty-four trailers and twenty-five tractors. This investment was followed in short–order by his purchase of a single–engine Cessna 210 that he learned to pilot, a truck yard in Reynosa, and a warehouse in Mission, Texas. He also employed a secretary, three salesmen in addition to himself, a traffic manager in Reynosa, and an employee at the Mission warehouse who supervised inventory and relayed orders telexed from the Houston office to the traffic manager in Reynosa.

With all of these assets in place Revelette devoted himself tirelessly to the goal, as one of his competitors, Bill Cypert, expressed it, "to get everything tied up, get it all tied up and let everybody buy their material from him, to control both production and prices." Cypert, who had preceded Revelette as a Great Southern salesman and subsequently shared some of his Mexican brick sources with him, considered Revelette to be "a big wheeler–dealer, a real hustler, a hard worker with a gift of gab who made a good impression, tall and nice–looking. Mexicans liked him, they like 'em flashy." Lester Tisdale, in the same vein, said: "Chuck was a go–getter and a hustler. He bought brick from anybody he could. And he did it his way. He forced 'em. He wanted repeat business." In reacting to Tisdale's comment [which I summarized for him], Revelette admitted, "I couldn't take 'no' for an answer. I went in to get something done."

Revelette's aggressive style with suppliers is illustrated by an anecdote he told me, chuckling as he talked, about his initial dealings with Ramón Longoria, owner of Camargo Brick in Ciudad Camargo, Tamaulipas. When I mentioned Longoria's name to Revelette in my 2008 interview with him, he interrupted and stated:

> Did Ramón tell you about me putting a casket full of brickbats on his front steps? He wouldn't sell to me, and we were having a pissing contest, and I figured that I was going to win. I bought a casket, took it to his house, filled it with brickbats, and laid it on his front steps. He got all upset. It was like one of those mafia deals because he knew damn good and well that we'd better do business.

The gambit worked and Revelette started buying from Camargo Brick. Indeed, another of his anecdotes pertains to that period: "I was up at Camargo and had a load of brick on a truck and the driver headed out toward the bridge. Damned if he didn't get bogged down in mud right up to the axles. We went and rounded up a hundred or so people, used ropes, and we walked that damn truck all the way to the bridge, just like Chinamen."

The self-confident, almost brash demeanor displayed in his approach to Longoria apparently typified Revelette's approach to competitors. Referring to brick dealers in Houston who sold mainly American–made brick he said: "I ran 'em off. The Mexicans had the product at the right price; it was always cheap. I was kinda like a Wal–Mart. People wanted to buy it from me because it was cheaper. I just kept selling the damn things. I had a pissing contest with [name of specific dealer] and I ran him out of the brick business straight into a beer distributorship. He made lots of money after that."

I first heard about Revelette in 1992 from R.H. Bond, one of the Houston area's pioneering and largest importers of Mexican handmade brick, as a "guy that got into the business in the late 1960s and became the largest." According to Bond:

> Revelette and I were kind of bucking heads but he reached the point where he was doing quite a bit more and I began to back off because I could see the handwriting on the wall. He tried to take over the border. One, he spoke fluent Spanish. Two, he'd start early and stay late. He was very forceful, a big guy and he spoke their language as good as they did across the border. He would go down there when brick was in great demand, and he'd just go into a plant and buy everything they had.

One reason why Revelette's reputation spread far and wide in the border brick industry, and why he was able to do business with nearly every important brick plant from Matamoros to Piedras Negras, was because of the mobility provided him by piloting his Cessna. As he expressed it: "I'd send one, two or three trucks to *Ladrillera Reynosa*, others to Camargo, Miguel Alemán, and so on. The thing about it is I had this damn airplane and I was a pilot. In the morning I'd be in Matamoros, at noon I'd be in Miguel Alemán, and in the evening I'd be in Piedras Negras. And I'd send all of my trucks back and forth."

Revelette would fly over brick plants he was buying from, where he had trailers parked ready to load, to check kiln activity and yard inventories. He also had his Mexican drivers keep track of plant inventories through on-site visitations to make certain that when kilns were unloaded "we were there." This was to avoid plant owners "making brick to sell to somebody else." During his best years in the early 1970s, Revelette bought regularly from fifteen to twenty plants, and became the major investor in *Ladrillera Ramírez* in Cd. Díaz Ordáz. In his highest volume year during that period, he imported 38 million handmade Mexican brick for the Houston market. Joe Bolin who was just entering the Houston brick business at that time, having turned down a job at *Metropolitan*

Brick, estimated that Revelette "controlled about 90 percent of the Mexican brick market" in Houston.

Revelette's investment in Artemio Ramirez's plant in Díaz Ordáz was motivated as much, if not more, by his need to Mexicanize the nominal ownership of his fleet of trucks, as it was to secure a steady brick supply source in a 'dog–eat–dog' market. As he explained:

> I had all of my trucks leased to a Mexican company, *Ladrillera Ramírez*, in which I had forty–nine percent ownership. U.S. Immigration took me to task two or three different times. They didn't want me hiring Mexicans to drive American trucks. I told them that my drivers were employees of *Ladrillera Ramírez*. They had me bring Artemio Ramírez to testify. I had to go get him, and he told them that those were his trucks.

The Tex–Mex plant was among those that did considerable business with Metropolitan Brick. Omar Zapata's recounted that business, and Revelette's *modus operandi*, in the following terms:

> In the early years of the boom, Revelette was one of our biggest customers. He was daring [*audaz*], a great big gringo, very, very daring [*muy, muy audaz*]. He began to run thirty trailers loaded with brick daily to Houston, at prices ten to fifteen dollars per truckload lower than the competition, below the prices of buyers from the same plant. In other words, he started to make thirty runs daily to Houston without earning anything. So, that's the way he eliminated competitors. And afterwards, when we no longer had many buyers, he lowered the price he would pay us because he was the only buyer. So, he put prices on our brick, and then he raised his selling price in Houston [*Nos bajó el precio a nosotros porque no más él nos compraba. Entonces, él nos puso el precio, y después lo subió al lado de allá*].

When I read Zapata's statement to him, Revelette expressed no disagreement with it and explained that he had done nothing other than follow the law of supply and demand: when sales dropped in Houston, he would simply lower his buying prices in Mexico.

Ironically, at the peak of his success in 1974, one of Revelette's salesmen—without checking with him first—unwittingly made a large sale and delivery of brick to a notoriously untrustworthy Houston apartment builder. Revelette attempted to get the eighty thousand dollar payment due and, finally, to retrieve his brick, but without success since to do so would have involved him in a shootout with several of the builder's armed guards. At around the same time, two of his truck drivers in Reynosa, who it turns out were also members of a local gang of narco–traffickers, were caught smuggling drugs in their brick loads. Despite Revelette's personal lack of involvement, this incident led to the closure and loss of his truck yard and office in Reynosa. So, he shut down Metropolitan Brick, sold all but five of his tractor–trailer rigs to pay off creditors, and ended up going into the Mexican trucking business with a friend in Matamoros to haul scrap steel from the port of Brownsville to recycling mills in San Luís Potosí. During this period, he also bought sheetrock from a plant in Puebla, Mexico, and warehoused it in Mission, Texas for ultimate sale in the southern Florida market.

In later years Revelette ended up with a Mexicanized fleet of refrigerated trucks that hauled produce from Mexico to U.S. markets. When I asked him if he ever had thoughts after 1974 about getting back into the brick business, he responded: "No. It was time to move on. Cost me a hell of a lot of money. So I just moved on, never looked back." He admitted, however, that he had become a millionaire in the brick business.

What factors motivate brick producers to get involved in transport? The case of José María Cantú is illustrative. He was born in Reynosa in 1953; his father was an agriculturist who also had a business in buying and reselling used iron and junk. After completing studies in a local business academy, he went to work for Raul Valera whose business interests included several brick plants. He learned the brick business by running Valera's plants for five years, and in 1976–1977 with ninety–thousand pesos raised from selling a small truck and his family car, he started supplying customers in Houston he had met when working for Valera with brick bought from small brick plants and *campos* for fifty dollars a thousand which he, in turn, resold them for fifty-five to sixty dollars a thousand. The arrangement with the small producers he described as "*refaccionar gente*" [i.e., providing financing to a brickyard operator to enable him or her to pay brickmakers]. He explained this as follows: "Clients loaned money to me—–two thousand, three thousand, five thousand dollars. With that money I began to make brick with people who had small kilns. I told them: 'You make brick. You tell me how much you spent on wages. I will give you what you spent on wages. I will pay for your firewood. You fire the brick and I will take them away'."

At first, Cantú used independent truckers from both Texas and Mexico to haul brick to his Houston customers. By 1978, however, he had rented a brick plant with wood–burning kilns and by 1979–80 leased a larger plant with gas–burning kilns. In 1981 he bought a four–hectare site, constructed kilns and installed a gas line. This plant was known as *Rey-Tam*. During these years, Cantú was shipping anywhere from twelve to thirty-five or more trailer loads of brick weekly to Houston.

Regarding the trucking question, Cantú emphasized that he got into trucking himself by necessity. He had fifteen to twenty trailers hauling for him, owned by several different truckers. These individuals met his suppliers when they loaded brick and met his clients in Houston upon delivery. In short, he began to see a pattern emerging as, one by one, some of the truckers who had hauled for him started buying brick from his suppliers and reselling them to his clients in Houston. To eliminate this spin-off competition, Cantú began to acquire his own trailers. He acquired five tractor–trailers of his own but also continued to use independent truckers who were not interested in directly competing with him. In his words: "I bought one and then another and two more. With the trailers that I put into it, I eliminated some truckers and others continued to help me. I had to cut some out. That's why I say that I was a brickmaker first and became a trucker only by necessity." After the brick market declined in the late 1980s, Cantú nationalized his fleet of trailers and started a freight hauling business in Mexico. When the export market began an upturn in the mid–1990s, however, he decided to set up another freight line in Hidalgo, Texas to once again haul brick to Houston.

Conclusions

The take–off into rapid growth of the artisanal brick industry on the Mexican-side of the border following the demise of that industry on the Texas side was a predictable outcome of a history of labor, capital, and cultural movements and linkages going back to the previous century. However, the amazing proliferation of artisanal brick production in Mexico during the 1960s and 1970s was not predictable on the basis of the prior history of cross–border movements and linkages because it was influenced by off–border forces and conditions.

There is no question that historically speaking the huge surge in growth of Mexican handmade brick production in the 1970s reflected under–capacity in the mechanized Texas brick industry in the face of an unprecedented boom in Texas masonry construction. However, there was also an aesthetic factor as well as a cost factor that underpinned the demand for Mexican handmade in Texas. Mexican handmade has a rustic southwestern look and its price per square foot on an exterior wall, reflecting its price per thousand, was substantially less expensive than that for U.S. machine–made brick. Still, it was Mexican handmade brick's availability in years of short supply of U.S. machine–made brick, in combination with its price advantage, that mostly explained its popularity among Texas builders.

The rapid expansion of Mexican gas–firing *ladrilleras* to meet the supply needs of the undersupplied Texas brick market in the 1960s and 1970s, was facilitated by their access to a deep labor pool that was replenished continuously by migration and recruitment of brickmakers from the interior of Mexico. This enabled the *ladrilleras* to expand their output exponentially through a satellite field/putting–out strategy which worked in two basic ways: 1) use available or newly acquired land and newly hired workers to start–up satellite brickfields to produced unfired brick that would be fired in the kilns at the main plant; or 2) sub–contract with smaller, wood–firing *ladrilleras* to produce and fire brick under the brand of the larger gas–firing plant.

The boom market of the 1970s was an anomaly in the history of the border brick industry, that was no stranger to market fluctuations, simply because Texas's major producers were unable to supply the surging market due to a lack of plant capacity. This left a significant share of the new Texas market to suppliers producing brick of all types in Mexico. The size of that share was destined to diminish when Texas plant capacity increased, a process requiring about five years, or when the surging demand decelerated. As it turned out, both of these conditions were present by 1980. Until then, even though there was 'dog–eat–dog' competition among Mexican brick suppliers, Omar Zapata could assert with confidence that handmade Mexican brick always found a market in Texas.

The steep decline in Mexican handmade brick exports to the U.S. after 1980 also reflected the organizational weaknesses of Mexican producers who could not agree upon, much less enforce, uniform product standards or marketing and promotional strategies. This left them defenseless to counter a sophisticated, multi–media negative advertising campaign sponsored by Texas brick producers through their industry association, the *Brick Institute of Texas*—a campaign best–known for the widely disseminated image of the president of that association taking a bite out of a Mexican handmade brick, juxtaposed with images of the exterior walls of homes built in cities like Houston with such brick severely damaged by rain or freeze–thaw cycles (Cook 1998: Ch. 12). In retrospect, a "few bad apples" line of defense invoked by Mexican producers does have some merit in explaining the undeniable fact that unfired, underfired, or misfired brick inevitably made their way through a chain of unscrupulous participants from Mexican plants to Houston walls.[6]

Notes

1. The location of Los Cavazos is shown in the middle of the far left side of Map 1. The Varela plant site is indicated by the black dot immediately below the Cavazos marker on the left margin of the Map.

2. *Rancho Las Calabazas* is located in the northeastern side of the landform it shares with Los Longoria and lends its name to the area in its immediate vicinity [see upper center portion of Map 1]. Its history and identity are overshadowed by that of *Hacienda Los Longoria* and I did not find a separate *Expediente Agrario* for it in the archives of the agrarian reform office in Reynosa.

3. Map 5 is a digital scan of a portion of a map produced by the United States Geological Survey in cooperation with the Department of the Treasury, U.S. Customs Service from photoimagery rectified by optical scanning from a color infared aerial photograh taken July 12, 1982 by the *Dirección General de Geografía del Territorio Nacional* (DGGTN), Mexico D.F., Mexico. Its tital is Reynosa, Texas-Tamaulipas (sheet 191, #26098-A3-PI-025).

There was another historically controversial cut–off downriver in the Mercedes area early in the century when the American Land and Irrigation Company interrupted a process of natural cut–off in the river channel at a hairpin bend and proceeded to make an artificial cut–off at the next horseshoe bend in the river so as to maintain the viability of an ongoing pumping station project. This cut–off changed the international boundary and arbitrarily changed the citizenship status of the Mexican–descent population of a local community named *El Horcón*. The citizenship status of these people was not resolved until 1970 through an international treaty (see Johnson, ed., 1991:100).

4. The origins of CORETT within the Mexican federal agrarian reform bureaucracy lie in the rapid demographic growth in rural Mexico in the 1940s that led to massive migration to urban areas unprepared to absorb the rural migrants. This led to a nationwide outbreak of invasions of periurban areas especially lands that were either *ejido* or communal in tenure, resulting in changes in patterns of land use from agriculture and ranching to illegal possession and commercialization for human settlement. These areas lacked basic urban services such as electricity, water, sewage, and so on, and became known officially as 'irregular human settlements' [*asentamientos humanos irregulares*]. To deal with the multiple challenges presented by these proliferating settlements, the Secretariat of Agrarian Reform established CORETT in 1973 within the *Departamento de Asuntos Agrarios y Colonización* to address these challenges, together with INDECO and *Fondo Nacional de Fomento Ejidal* (www.corett.gob.mx).

5. My interview with this informant took place in my car parked in front of a mobile home on an unpaved farm road north of Weslaco, Texas. González, who lives in Anzaldúas near "Boca del Rio' north of Reynosa City on the riverine highway, was visiting his son at the time of the interview. I found out about him from Jovita Herrera, daughter of Manuel Herrera who was the foreman at the Relámpago plant where Jovita spent a lot of time as a young girl. She is married to one of Luciano's brothers.

6. A personal communication to me dated July 12, 2004 from D.L. Langford of Houston provided yet another first–hand account of the hand– and machine–made brick business from the late 1960s to the end of the 1970s when he worked for the Granger Company, a large Houston distributor and competitor of Revelette's Metropolitan Brick. Langford considers Ramón Longoria to have been "the most intelligent and resourceful of all the brickmakers along the Frontera . . . who kept many of the other brickmakers afloat" with "the biggest operation." In his opinion: "Mr. Longoria was right when he said that 'any damn fool can make handmade brick'. I think I know what he meant. He meant that the mixing is no big deal. I can vouch for that since I tried it a few times myself. The art is in the burning . . . machine–made brick was the real art." According to Langford, the Granger Company "sold over 10 million brick a month in the Houston area, and we dealt with brickmakers from Piedras Negras to Matamoros to East Texas and Louisiana."

Chapter Six
Class Tensions, Collective Bargaining, and Business Strategies

In countries like Mexico, where capitalism is regionally and sectorally differentiated, handmade brick is one of those commodities that lends itself to production in enterprises where workers are employed through a piecerate wage. Even in Mexican regional economies like that of Oaxaca with long traditions of household– and village–based agro–artisanal forms of production, full–time brickmaking typically occurs in small–scale enterprises organized through an employer/pieceworker relationship (Cook 1984:124–130). Although the nature of piecework is such that it provides the individual pieceworker leeway to work at his own pace and timetable with a minimum of supervision, there is still an ongoing engagement between employer [*patrón*] and pieceworker focused on remuneration that generates friction. This friction may lead to confrontation and conflict if it is not resolved through some form of negotiation or collective bargaining process. In Oaxaca, such friction tended to be dissipated by employers ultimately granting raises to pieceworkers, so as to avoid losing them to a competitor. This tended to lessen the pressures toward the emergence of collective bargaining but did not keep workers from changing employers to improve their remuneration (Cook 1984:139–148). By contrast, under conditions of a burgeoning export market and hyper–expansion of the handmade brick industry on the Mexican side of the lower Rio Grande border in the 1960s and 1970s, the contest for wages and profits became openly antagonistic, at times erupting as strikes or lockouts requiring resolution through collective bargaining.

The Contours of Labor Relations and Discourse

Historically, employers' associations in the lower border brick industry have used the terms *productor* [*producer*] or *fabricante* [*manufacturer*] in their official names, whereas the workers' union used neither of these terms, and was known as the *Only Union of Workers of the Brick Industry* [*Sindicato Unico de Trabajadores de la Industria Ladrillera*]. Unofficially, however, members of the producers' Association or the Union were referred to individually or collectively by others and themselves as *ladrilleros* which simply implied an affiliation with a *ladrillera* and the brick industry.

Employer discourse exposes contradictory tendencies regarding workers in general: one that devalues brickmaking labor as unskilled, and another that recognizes its value and unique artisanal quality. Consequently, some employers contend that anyone can make brick by hand. For example, Ramón Longoria, introduced in chapter 4 as a prominent mechanized *ladrillera* owner who also managed handmade brick production, explained that he "came to the conclusion that anybody could make handmade brick but not everybody could make machine–made brick." By contrast, Joél Rodríguez, owner of *Ladrillera Reynosa*, in reminiscing about his long experience in the industry observed that "not everyone can make brick" and emphasized that brickmaking had to be taught and learned.

It is instructive to give some background on Longoria's position. He grew up on his father's ranch in Reynosa but, before 1900, his father and uncles also had property on the Texas side of the river south of Madero. They had kilns and made brick for their own use there. Longoria insisted that "you could go out anywhere along the river on either side, get yourself a hoe, and make yourself brick. And then you dry and stack the brick that you make, put mud all around it, leaving some openings for the heat to escape and the air circulate, and you burn it." He also recalled a meeting on the Mexican side of the border back in the 1960s involving plant owners where he and others who had mechanized plants kidded the rest by asserting that "any fool can make handmade brick" [*Cualquier pendejo puede hacer ladrillo a mano*]. He elaborated on this theme as follows (1993 BBP Edinburg Roundtable transcript):

> Anybody can make handmade brick, especially if you don't know you can't do it. All you need is a shovel, a place where you can get some water, and a mold, get some firewood, make brick and burn it. The thing is that when you make handmade brick, it's easier to dry because you use a lot of water and it's sandier soil, it doesn't have so much clay. And then, once you put it into the kiln, you can't over fire it because you've got to put more wood in the kiln. That kind of controls it so you won't go to the excess too fast. If you put enough wood in there you are going to get as good a brick as if you had fired it with gas. So anybody can make brick, even I made brick.

Longoria, like Guenther Weiske, was drawn toward mechanized production, and succeeded in building his large Camargo plant into a mostly mechanized operation—but not without experiencing significant problems with the process, especially the continuous kiln component, prior to selling out to the Gustavo García Garza in the 1980s.

Although mechanized production brought its own set of problems, there was near unanimity among the largest handmade plant operators that their workers, especially during the boom years of the 1970s, tended to be unreliable. According to René Ramírez, long–time foreman at *Ladrillera Reynosa*: "Brickmakers usually have a little problem of liking beer too much, especially on weekends, and for that reason on Mondays they don't work." Omar Zapata, who had his own wood–burning plant but also worked as a production manager at *Ladrillera Reynosa* and at *Tex–Mex* stated: "People are unreliable in Mexico. Sometimes they show up to work, other times they don't. It's hard to predict. At the brick plant, workers don't come at 8 a.m. like they do in other businesses.

They can come to work or not; missing Monday and Tuesday and showing up on Wednesday. I could never be sure if I would have seventy or one hundred workers on any given day." Still, Zapata, like every other handmade–brick businessman, realized that dependence upon the human labor element was inherent in their business, and that the ability to recruit and maintain brickmakers was the key to success in the industry.

For Zapata, the main way by which these special businesses recruited and retained labor was by paying workers at a rate higher than minimum wage, typically as much as one–third higher. "It has to be higher," he asserted, "because even though it is rustic work you have to know how to do it. To know how to mix clay, to mold it, and so on. That is, to be a brickmaker, to be able to do it. Not everybody can do it."

But even when brick industry workers were characterized by employers as prone to unreliability, they were also viewed as being opportunistic, in the sense that cleaner or less physically punishing work was seen by them as being more attractive and drew them away from brickmaking. For this reason, according to Zapata, it would be difficult to rehire workers today who had previously worked in the brick industry: "At present I consider the labor supply to be problematic. Workers who have had easier jobs than brickmaking won't be inclined to work as brickmakers again. It is very hard work." In short, Zapata's twenty-one years of experience as a brick plant manager convinced him that dealing with workers and addressing their problems was the most difficult issue confronting the industry.

A principal expectation of the workers was free housing and utilities which, from Zapata's perspective, simply meant exacerbating the labor management problem. In his words:

> It was relatively easy to attract workers. You offer them a little room with electricity and running water. But if the worker gets mad at the company and starts causing problems, it is difficult to get rid of him [Mexican law makes it difficult to dislodge recalcitrant workers or to cut off their utilities]. On balance, its better to have workers commute to and from the plant daily. It is better business not to have resident workers, not to have worker housing.

Rather than provide housing with utilities as many plants did, Zapata at Tex–Mex hired commuters and provided them with transportation to and from the plant. For example, he hired ten or twelve workers from *Colonia El Banco* which, he correctly notes was very, very far away from the plant, and paid weekly for a *pesera* or public car to take them to and from the plant. At least one large Reynosa area brickyard, the Varela's *Ladrillera Industrializada Rio Bravo S.A.* which was located in Cavazos not far from the Tex–Mex plant, had shuttle bus service for its non–resident workforce. Aside from transportation, Zapata provided his Tex–Mex workers with several well–maintained bathrooms and showers, as well as a lunchroom where a cook prepared daily a lunch of staples [rice, beans, noodles, tortillas].

Employers' views devaluing brickmaking labor and complaints about the unreliability of the labor force are, not surprisingly, strongly contested in worker discourse. Many workers insisted that their employers were uninformed about brickmaking, and wholly dependent upon their employees to correctly perform all of the tasks required to proper handmade brick. The most articulate spokesman for this viewpoint was Manuel Lozano who came to Reynosa at the age of fifteen in 1935 from a rural area of Nuevo León. He worked in the plant of Alberto Gómez in La Copa for five years, and then spent several years as an agricultural worker in the Rio Grande Valley before returning by 1948 to work in several Reynosa brick plants and fields. During the peak years of the union's effort to organize brick industry workers, Lozano was an active supporter. In reminiscing about his decades of work in the industry he expressed the opinion, with reference to the owners of several semi–mechanized plants where he worked: "They didn't know anything about brickmaking. They just didn't have a clue. We were the ones who told them how they could control and make the material. There's more to being a brickmaker that just saying 'I am going to make a brick.' Not every man can make brick."

In the same militant spirit, Lozano who was at the time foreman of the brick molding area [*mayordomo de los patios*] in Juan Tueme's plant said that he, and some of his co–workers, chided Tueme one day regarding his lack of knowledge of brickmaking: "You do not understand the work involved. You just don't understand. You need us to tell you how to do it. You may have the money but lack the knowledge. The work, we know how to do that." Tueme, according to Lozano, admitted, "No, I am not a brickmaker." The bottom line for Lozano was that, with few exceptions, plant owners knew how to account for brick made but not how to make them: "We made brick, they made the numbers." He added that one day Tueme confided to him that in the first year of his plant's operation he had lost one hundred and eighty thousand pesos [fifteen thousand dollars] due to his lack of knowledge of how to make brick. Lozano attributed this loss, if true, not only to a lack of knowledge but to poor organization and to a lack of roofed molding and drying sheds essential to minimize losses due to rain. Tueme later added these at his two large gas–firing plants.

What is the background profile of a representative cross–section of twelve Reynosa–area piecerate brickmakers who were employed during the decades of "brick fever"? A majority were born in rural areas in the interior of border states of Tamaulipas, Nuevo León, or Coahuila and came to the border area as adolescents or young adults with no prior work experience in brickmaking. A minority were born in rural areas of non–border states including Zacatecas, Durango, San Luís Potosí, Guanajuato and Querétaro. With a few exceptions, most of those who were raised in the Reynosa area were not born there, and their earliest work experience was in ranching or agriculture rather than in brickmaking. None of the brickmakers had more than three years of formal schooling and most of them had no formal schooling. The average age of the sample is fifty with the range from thirty to seventy three; the average number of years of work experience in brickmaking is twenty–eight. Approximately two–thirds of these workers also had some work experience in the U.S., mostly in the agricultural fields of the Rio Grande Valley. More than half of the sample over fifty years of age also spent some time as undocumented immigrant brickmakers on the Texas side of the corridor.[1]

It is a mistake to attempt to analyze work patterns of piecerate brick molders strictly in terms of a pecuniary cost–benefit logic for two reasons. First, there is definitely an upper limit on physical productivity per day and per week in brickmaking which, in turn, limits the capacity of pieceworkers to respond to material incentives to increase output. Second, a significant number of pieceworkers, from the perspective of their capitalist employers, own testimonies, and my observations, display a propensity to opt for leisure, as opposed to additional work, once a target income was reached. Many brickmakers, for example, tend to work a four–day week, reserving Saturday, Sunday, and Monday for rest and relaxation, a period that traditionally included high levels of alcoholic consumption. Admittedly, there is no systematic observational or survey data to support this generalization and it may be more applicable to the boom years of high employment than it was afterwards when the workforce was reduced to career brickmakers.

The following statement by Joél Rodríguez, elicited by my question regarding wage incentive schemes, is more revealing than most but otherwise typifies capitalist discourse regarding pieceworker behavior:

> A wage incentive scheme was always given to them. But they are very special people. Since they are artisans, they work how they want and when they want. Those people and their idiosyncrasies—it's simply that they want to cover their basic needs, and once they have covered them, they have no incentive to work. In general, our people unfortunately have lots of vices and prejudices. That is, independently of the fact that incentives were offered, on a piecerate basis, bonuses were given, these had no great effect, no results. Why? Because those people were saying 'I need X number of pesos to eat and so many pesos to get drunk and since I've already earned them I'm going to work only 3 days this week. These are people who have never worked a whole week, never. They work 2 or 3 days, 4 days, and then stop working.

René Ramírez, in response to my question regarding how his brickmakers reacted to wage incentive schemes in the past, replied: "They were not interested in earning more. Every raise that was given to them—for example, if they were making one peso per brick and were given a raise to 1.20 pesos if they made 1,000 or more brick, they would produce only 950 because at the rate of one peso they earned enough." When a clay mixing machine was introduced in this *ladrillera*, thus relieving brickmakers of the task of preparing clay, the 1,000 brick per day quota was then realized by many pieceworkers but, rather than producing an additional lot of brick in the time saved by releasing them from clay preparation, these molders opted instead, according to Ramírez, to quit early. In his words, "The people finished sooner, that is all they wanted. The only benefit that came from the mechanical clay mixer [*batidora*] was that all the clay was uniform."

Ideally, in the handmade brick industry productivity increases can be achieved by adding pieceworkers so long as there is a pool of willing and able workers. When labor is scarce, selective introduction of labor–saving machinery may achieve the same results and also increase worker efficiency as, for example, was the result of introducing mechanical clay mixers that eliminated the need for molders to prepare their own clay batches. Chapter 9 includes a detailed discussion of this process, and its results, at the new *Ladrillera Reynoso*.

Worker discourse both contradicts and reinforces capitalist discourse regarding work patterns. One molder who characterized himself as "*ambulante*" [itinerant], and who worked two, three or four months yearly in the brickyards and then moved to construction, made twenty–five hundred brick over a four–day week in 1993 to earn seventy five New Pesos [twenty–five dollars] which he said was "to eat, nothing more" and added "If I have food, it's o.k." When I asked him why he didn't make more brick daily to increase his weekly income, he replied, "It's hard to make a thousand brick in one day, one struggles a lot to do it and the next day one is tired out; I wake up the next day and can't go to work."

Manuel Lozano, the seventy–three year old retired brickmaker quoted above, emphasized that although many pieceworkers made relatively good money in the 1960s and 1970s they did so " . . . not because the boss paid very well but because you made more than the limit, you forced yourself to make material. That's the reason. Because you forced yourself to get the money you needed, not because the boss rewarded your work." But, he admitted, "If you produced more, at the end of the week you got a reward . . . so people would exert themselves more." His statement implies the effectiveness of employer incentive schemes. Lozano also noted that "others produced more because they told themselves 'I need such and such an amount of money' to which the *patrón* responded 'Well, get to work mixing the clay.'" Lozano also admitted that he personally "never forced himself to make a thousand or fifteen hundred brick a day" and was of the opinion that those who did so "are no longer with us." He chose not to accelerate output because "work is not like food. Work must be sized up. The more you do, the more you need and the more effort you make, well, you end up making life more of a drag" [*El trabajo no es alimento. El trabajo hay que tantearlo. Entre más se hace, más se necesita y más se esfuerza. Pues, uno hace la vida pesada*].

To sum up, brickmakers are a mixed bag: some respond to piecerate incentive schemes and work harder and longer to produce and earn more, whereas others march to their own tune in the form of a targeted income or package of wage–goods, typically coordinated with a leisure time allocation that constrains self–exploitation (see Cook 1995:84–87 and 2001).

Labor Relations and the Brickmakers' Union

Historically, workers at some of the largest gas–firing brick plants did participate in unionization efforts that began in the 1950s and continued with mixed results into the 1970s. Not coincidentally, the take–off in the Reynosa brick industry after 1950 paralleled a growth spurt in the government–run petroleum industry, *Petroleos Mexicanos* or PEMEX. The discovery of gas fields in the region spurred the expansion of PEMEX infrastructure, including offices, housing, a refinery and a petrochemical plant, and made the Petroleum Workers Union [*Sindicato de Trabajadores Petroleros de la República Mexicana* or STPRM] an important player in the political economy of Tamaulipas.

As outlined in chapter 4, the construction of a natural gas line across the riverine zone of the state of Tamaulipas enabled the expansion and technological upgrading of the regional brick industry. Given this co–development of two industrial sectors, petroleum and brick, it is not surprising that the first attempts at organizing brick industry workers apparently came from the Petroleum Workers Union. Nevertheless, the process was mostly identified with the *Sindicato Unico de Trabajadores de la Industria Ladrillera*, affiliated with the CTM [*Confederación de Trabajadores Mexicanos* or Mexican Workers Confederation], that operated at the municipal level in Reynosa and Miguel Alemán. Conflicts emerging from the collective bargaining process in the border brick industry were, according to Mexican federal labor laws, presented for arbitration at the local office of a federal agency, the *Junta Federal de Conciliación y Arbitraje* [Federal Arbitration Board].[2]

The record of the brick industry labor movement over the decades is uneven. It is sketchy for Reynosa where apparently only two plants were successfully unionized for a short time, but more complete for Miguel Alemán where records are available from the 1970s for one of the plants unionized there. Plant shutdowns—due to strikes, lockouts, or in at least one case in Miguel Alemán, worker occupations—occurred. Nevertheless, it is clear that many large, gas–firing, export plants in Reynosa managed to avoid disruptions attributable to labor–management conflicts caused directly by union activity.

The decade of the 1960s presented the large semi–mechanized, gas–firing plants with another problem besides export licensing discussed in chapter 4, namely, labor unrest and the challenge of unionization. Addressing this problem required inter–plant coordination and had the effect of strengthening the Association: In the mid–1960s Salinas, together with several other owners of major *ladrilleras* including Juan Tueme, Raul Valera, Gustavo González, and Gustavo

García were confronted with an aggressive unionization effort apparently spearheaded by the petroleum worker's union. A strike was called that lasted for several months but was successfully resisted by the owners of the major *ladrilleras* in the Reynosa area—not one of which was successfully unionized. The major successes of the unionization effort were restricted to the Camargo–Miguel Alemán area.

Salinas made the following statement about what he called "this very difficult period":

> When the business grew . . . there were people interested in forming a union. The leader of the Union of Petroleum Workers became interested in organizing a brickmakers' union. His interest was apart from the benefits that would be gained from organizing the brickmakers. They were strange interests that were neither to the benefit of the union nor the businesses. Interests to escalate political things, to form political groups to obtain benefits.

Given the unevenness of data regarding this period of labor unrest, we are left with Salinas' somewhat cryptic interpretation, and with the more important fact that many of owners successfully resisted a strike and the unionization effort. "We had a very serious strike that lasted four or five months," Salinas explained. "In order not to accept the union we suspended production. With a union we were not going to be able to remain viable. We succeeded in keeping the union out. We won the strike. There were only a few enterprises where they had a union. Don Ramon Longoria always worked with the union." Despite the concerns voiced by Salinas, in those plants where the union was able to organize [e.g., *Ladrillerías Camargo*, *Ladrillera Terra–Mex* and *Ladrillera Colonial* in Cd. Miguel Alemán, and *Ladrillera Tueme* in Reynosa] available data show that it was effective in negotiating contracts that benefited workers without making the unionized enterprises noncompetitive.

Despite the efforts of government–affiliated unions, labor–management relations in many enterprises in Reynosa's brick industry seem either to have been non–unionized or governed by so–called 'white unionism' [*sindicalismo blanco*] that is by workers' associations or groupings not strictly within the orbit of the CTM. Mireya Ramírez Acosta, managing owner of the Terra–Mex plant in Miguel Alemán, described this type of unionism as an antidote to the militant CTM regime that she confronted:

> Workers were unionized but did not belong to the CTM. The workers organized a group and made an agreement with their employer. The enterprise provided them with benefits and financial assistance according to the Federal Labor Laws [*Ley Federal de Trabajo*]. Workers did not strike as the CTM did. They could have problems with the enterprise and they addressed them across the table in a meeting without having to go to arbitration. That is, without having the participation of the CTM. Everything was arranged internally.

Nevertheless, in the 1960s in Reynosa, the CTM version of unionism and so–called *white unionism* were in a struggle to control labor–management relations in the brick industry.

Manuel Lozano was working in the plant of Gustavo García in Reynosa one day in the early 1960s when Garcia assembled his workers. He surprised them with the news that starting on the following Monday they would be working under a twenty–eight–day contract that would have to be renewed, and signed by each worker, every subsequent twenty eight–day period. This was apparently an effort by García to preempt a process of negotiating a contract under CTM direction. Manuel and several other workers resisted the idea on the basis that it demonstrated García's lack of confidence in them. When they expressed this position to García he reportedly presented them with an ultimatum: "If you want to continue to work here you sign the contract, if you don't sign you won't work here." Lozano said that he responded to Garcia on behalf of the nonconforming workers as follows: "Don Gustavo, don't be that way with us. We taught you how to produce brick. We are doing the work, and you are putting the price on our work. What more do you want from us?"

There was a standoff between García and his workers. Lozano took the initiative in organizing his co–workers to present García with their own version of a contract that would be mutually agreeable, presumably one drafted through consultation with union or CTM representatives. García continued operating his plant with workers unsympathetic to the union but Lozano and twelve others who had worked for García began to circulate around other plants, each of which had a designated union delegate. All told, there were some one hundred workers from various plants who were sympathetic to the union cause. Lozano referred to the union as "*el Sindicato de la Industria Ladrillera*" [a shortened reference to the CTM affiliate *Sindicato Unico de Trabajadores de la Industria Ladrillera*], and characterized it as operating strictly at the municipal level. There were approximately seven plants with union delegates and supporting workers, including those of Alberto Gómez, Efraín Salinas, Gustavo García, Juan Tueme, as well as *Ladillera Mexicana* [LAMEX] and *Ladrillera Morelos.*

Although Lozano and other's approach to employers for wage increases was assertive, antagonism was muted. They did not consider their efforts to be contrary to those of their bosses but simply as reflecting their right to pursue a living wage at their place of employment. "I knew," explained Lozano, "that the enterprise came first because without an enterprise you don't eat—you don't eat, I don't eat, the foreman doesn't eat, nobody eats." He admitted that "we scared them" but reiterated that "the purpose of us having organized ourselves was not to harm the enterprise but rather to maintain it because the employer, the foreman and all the union workers eat from it."

A countervailing reality motivating their movement, however, was according to Lozano that "the bosses did not want to pay us very well." Indeed, he claimed that a wage of thirty pesos [$2.40] per thousand brick had prevailed for thirteen years during the 1950s and 1960s, and the workers joined together to ask for a modest wage increase to thirty–five pesos [$2.80] in 1970. Regarding this situation Lozano commented retrospectively: "Isn't the boss going to notice that the price of potatoes is going up? What consciousness! It's not that you don't want to work, it's that you need the income. You need to take something home. That was the problem."

Juan Tueme was one of the few employers in Reynosa who settled with the union. The majority of plant owners choose not to negotiate with the union representatives and responded with a lockout that lasted for several months. The union was never able to organize industry–wide. Nevertheless, no plant could operate without meeting the labor standards set forth in federal labor statutes or without conceding wage and non–wage benefits negotiated in the region at unionized plants. Not surprisingly, it was around this period that many of the same Reynosa plants resisting unionization began to actively recruit brickmakers from the interior of the states of Nuevo León, especially the Monterrey area, and Coahuila [Torreón and the *ejidos* of the Laguna area]. The establishment of fields outside the precinct of the main plant, ostensibly to permit output increases beyond the capacity of the main plant, also had the effect of reducing the likelihood of unionization, controlling wages, and eliminating social security payments.[3]

The available data suggest that the union regime in the Díaz Ordáz–Miguel Alemán–Camargo region [see Map 1] was more aggressive and longer–lasting than in the Reynosa area. José Angel Gutiérrez, as head of the *Federación de Trabajadores de la Zona Ribereña* located in Díaz Ordáz—the regional office of the CTM——was an aggressive union organizer and leader who played a particularly strong role in maintaining a union regime and resolving collective bargaining disputes in that area. This reflects his prominence in the regional power structure dominated by the government party, the *Partido Revolucionario Institucional* [PRI]. Reportedly no business could operate in Miguel Alemán—restaurants, construction, brickmaking——without having its employees belong to a CTM–affiliated union.

The Terra–Mex plant in Los Guerra, near Ciudad Miguel Alemán, provides insight into the details and dynamics of the unionization of *ladrillera* workers (see Cook 1998:Ch. 9). The workers were unionized by the *Sindicato Unico de Trabajadores de la Industria Ladrillera y Conexos del Municipio de Miguel Alemán*, affiliated with the CTM, and worked for the firm in accordance with a collective work contract [*contrato colectivo de trabajo*] that stipulated working conditions and - rates. Despite this union–sanctioned contractual relationship there was still a decidedly paternalistic character to employer/employee relations and to daily life at the plant.

Mireya Ramírez of *Terra–Mex*, viewed the union membership of her male workforce as a factor that contributed to resistance against her role as *patrona*. As she expressed it: "It was a very difficult experience since these men belonged to the CTM that was . . . very strong and we had serious labor problems." She elaborated:

> The workers measured the limits of my capacity to give orders, if I was too flexible, or too hard. I tried to be hard when necessary and to be flexible when necessary. That led to good understanding and agreement. I withdrew when we had labor problems. There were several employees, would–be leaders, who were harmful to the interests of other workers. I was obliged to fire them regardless of the consequences. I got rid of those workers who caused unrest. They were just self–serving. They were so brazen that on one occasion they confronted me and said: 'Look, it's in your interest to treat us well and we'll keep the rest in line'. That is when I fired them. There were four or five of them who were agitators. They took me before the conciliation board but I got rid of them. Afterwards, the business operated smoothly. I followed the guidelines of the Federal Labor Law and everything was peaceful.

The union engaged in some strong–arm tactics at the plant to enlist workers and to maintain discipline. During the late 1950s when it began organizing brick industry workers in the Miguel Alemán area, Antonio Muñatones, the *Terra–Mex* plant foreman who became a union member due to intimidation, described the situation as follows: "The union organizers wanted all of the workers to join the union and since our most reliable workers were not unionized, they were pressured." He was approached by a union organizer who told him, "If you do not cooperate in organizing a union, we will beat you up when you come outside." That threat led him to talk with the plant owner about the situation, and the plant owner, who was Mireya's brother, told him to go ahead and join the union. Antonio did so grudgingly but for several years thereafter paid his union dues weekly and attended biweekly meetings in Miguel Alemán.[4]

Ramirez complained about this and other union practices in retrospect: "Every two weeks they stopped working because they had to go to a union meeting. They had to attend since they were fined if they didn't. And they had to vote for PRI." Muñatones acknowledged this practice by saying, "We had to attend." [*Teníamos que acudir*]. Ramírez proceeded to give her own version of union strong–arm tactics:

> It was like a threat. The union instilled fear in them. If they didn't participate they would do this or that to them, that they would leave them without work, that they would make it difficult for them to work. They did not permit the employer to be present when they were talking to workers. They came and started talking and said that if the employer was present workers would be intimidated from speaking their minds. They came right into the plant where the workers were at work.

Despite her hostility and combativeness, Ramírez insisted that she was able to deal with the union without damaging the enterprise or her cordial relations with the majority of her employees.

A perusal of the collective work contracts reveals a content that fairly represents employer as well as employee interests not only in general but with a degree of specificity reflecting detailed knowledge of and sensitivity to the particular problems of border brickmaking. The fourth and fifth clauses of both the 1970 and 1974 contracts illustrate this:

> Clause 4: Enterprise and Union agree: to the extent that the lack of raw material (clay), rains, overproduction, breakdown of the water pump, breakdown of transport vehicles, closure of the border to the export of brick or American import duties prohibitive to Mexican brick, lack of a market, etc., will be motive for suspending work, without responsibility to the Enterprise, starting up again as soon as such circumstances disappear; the enterprise agrees to give aid to workers for six (6) weeks with credit corresponding to three work days per week according to the minimum wage for this zone, to each one of the unionized workers and upon resuming normal work the amount of aid given will be discounted proportionately each week
>
> Clause 5: Enterprise and Union agree: that in cases of rain or other meteorological phenomenon causing damage to brick made during the day, 50% of their total value will be credited; but when for reasons of worker negligence in not placing finished brick in the drying shed, if these suffer damage the Enterprise is not responsible.

The inclusion of the above clauses in brick industry contracts demonstrates the vulnerability of handmade brick production in the border region to inclement weather, even where capital had been invested in *secaderos*, that is, roofed sheds where molding is performed and brick are dried prior to firing. This vulnerability is much greater in the multitude of smaller–scale, less highly capitalized *ladrilleras* like those of Tano Arriaga and Anita Soto without *secaderos* where plastic sheets are used to protect undried and/or unfired brick from the rain. In the latter situations, of course, workers are not compensated for brick that are damaged.

A comparison of the wage scales [*tabuladores*] in the collective bargaining contracts shows a substantial across the board increase in wage rates between 1970 and 1974. For example, the rate for molding standard 3 x 3 x 9 inch brick went from 27.50 pesos [$2.20] per thousand in 1970 to 49.50 pesos [$2.90] per thousand in 1974; the rate for setting unfired brick in the kiln went from 4 pesos [$.32] per thousand to 10 pesos [$.59] per thousand; and the rate for drawing fired brick from the kiln went from 6.80 pesos [$.54] to 13 pesos [$.76] per thousand. It could be argued that this wage increase, as well as other items in the contracts, reflected the role of the union in looking out for worker interests—especially the bottom line of maintaining their purchasing power and standard of living under inflationary conditions. This may be the case in those areas like Miguel Alemán where a relatively high percentage of brick industry labor was unionized but in the Reynosa area, where most brick industry labor was categorized as "free" [*libre*] wage rates also rose periodically simply as a consequence of the dovetailing of the class struggle and the labor market.[5]

In short, brick industry labor did not need either government or independent unions to negotiate on their behalf; loose, ad hoc, self–organization by workers at discrete brick plants sufficed to enable them to periodically negotiate their interests with employers—although the threat of unionization probably made employers more predisposed to accommodate worker demands.

That the periodic problems with unionized workers at Terra–Mex were not unusual in Miguel Alemán at the time is illustrated by the first experience in the brick business of José María Cantú from Reynosa who for five years in the early 1970s ran the *Ladrillera Colonial* in Miguel Alemán for it's new owner, Raúl Valera. On the day that he took over the plant from its previous owner, Juan Angel García, the latter told him: "Well, I'm leaving this time bomb with the fuse lit in your hands. Let's see if you can deal with these people. I no longer can."

Cantú explained that the problem hinged on the wage scale. The workers demanded across the board increases that would have made the business unprofitable. They asked for wages that were comparable to the plant's selling price for brick. Cantú met with José Angel Gutiérrez, the head of the union for the riverine zone located in Díaz Ordáz, and showed him the plant's cost and price data. Gutiérrez agreed that the workers' demands were unreasonable and would force the plant to go out of business. The state Conciliation and Arbitration Council took the same position, and named a new board of directors for the union consisting of workers who were willing to accept a compromise pay scale. However, fifteen discontented workers who supported the higher pay scale rebelled and took possession of the plant, locking out the owners, managers and the thirty workers who accepted the compromise pay scale. This led to an intervention by state authorities who removed and arrested the protesting workers. Afterwards, the plant resumed normal operations for several years.

Despite the overall pattern of irregularity or periodicity in brickmaking employment, there are some cases of relatively long–term relations between pieceworkers and employers. Economics rather than sentiment are operative here—either the employer is willing to extend credit to workers or to keep them producing and earning, albeit at reduced levels, during market slumps; and/or provides them with rent–free housing, water, electricity and maintains their social security payments which enables them to get health care. Employers who do not offer these benefits, or who fail to meet payrolls, have difficulty in keeping or recruiting brickmakers from the local labor pool.

To reiterate, many of Reynosa's gas–firing brick plants were never incorporated into the union regime. After most of the large plants moved from the city center to the Rancho Grande and Cavazos *ejidos*, it appears that *ejido* authorities took the position that unions were not permitted on federal land. This was the case, for example, with *Ladrillera Reynosa*. According to René Ramirez, plant foreman, "When the union came around here to try to impose itself it was not able to because the *ejido* authorities did not permit it operate since this is a federal zone."

Nevertheless, the specter of unionization haunted Omar Zapata at the time he and Antonio Méndez, foreman, were opening the Tex–Mex plant in the mid–1970s. In the context of describing how Tex–Mex workers were not provided with housing but with restrooms, showers, and a lunchroom, he said that the only "ordinary, common thing [*lo único común y corriente*] was the union that I fought, and fought, and fought so that it could not get established." His strategy to avoid union problems was to "hire workers he knew or had confidence in" [*ocupar pura gente conocida, de confianza para no tener problemas*]. The realization of this strategy was delegated to Méndez who explained the meaning of hiring "known" workers:

> When we entered to work here at this plant, I knew everyone. I had dedicated my career to brickmaking and experienced difficulties caused by the union. Zapata told me that he wanted to work freely, without problems, and that he wanted to know everyone who worked for him. I worked for Salinas, with the Arab Tueme, in *Ladrillera Juarez* in Rio Bravo so I knew all of them.

Ironically, labor problems at *Ladrillera Juárez* in Rio Bravo forced Méndez back to Reynosa where he worked for a time during the mid–1970s in a *tortilla* factory since many brick plants were either on strike or shutdown to avoid confrontations with the union. Brick industry workers, including himself, in those months were circulating between plants and were divided over the union problem.

Mendez's position regarding the union was that people who joined it wanted to boss other workers around or threaten them. He also contended that union workers looked for any pretext to strike, and to force non–union workers who wanted to work to stop working. In short, as a foreman, Méndez clearly sided with management and, in this case, with his own boss, Omar Zapata, in opposing unionization of brick industry workers.[6]

Costs, Prices and Profits

At one point in my interview with him, Omar Zapata asserted with regard to the handmade brick industry that "it is a fabulous business. It's as good as it gets. It's among those businesses that leaves the best returns." He supported this assertion by claiming that during the prolonged period of high demand in the 1960s and 1970s they could sell brick at any price they wanted. The simple fact was that brick sold, regardless of whether the price was forty dollars a thousand, sixty dollars a thousand, or seventy dollars a thousand. Brick sold, there was always a market for them. Competition among Mexican sellers did not matter because everyone was selling, even if there were price differentials on the order of those just mentioned. There was no real competition from U.S. manufacturers simply because machine–made brick were much more highly priced than Mexican handmade brick during the boom years. Price differentials between competing Mexican sellers were relatively insignificant because of the high profit margins all sellers were experiencing. According to Zapata, differences of ten dollars per thousand could just have easily been one hundred dollars per thousand "because brick left a fabulous margin of return." He remembered that prices when he first came into the industry in the mid–1960s were around thirty–two dollars per thousand. When he retired twenty years later prices were around eighty to ninety dollars per thousand, and at some points in between prices rose to as high as $130 per thousand. But whatever the price, according to him, the profit margin remained more or less steady at 60 to 70 percent.

Zapata acknowledged, however, that even though returns to producers were "fabulous" they were even more so for intermediaries and builders. Therefore, the best possible situation—guaranteeing the highest returns over a prolonged period—was found in vertically integrated enterprises like Tex–Mex that produced brick in Mexico, and then built houses or apartments with them in places like Houston.

One variable Zapata did not factor into his profit equation was the dual currency conditions under which the border brick industry was operating, namely, that production costs in Mexico were paid in pesos, whereas export sales brought in revenues in dollars. Typically this resulted in potentially large windfalls for Mexican producers, facilitated by cycles of overvaluation and devaluation in an exchange market with shifting mixes of controlled and free–floating peso–to–dollar rates.

These conditions became even more challenging in the 1980s as peso–to–dollar exchange rates moved from twenty–three to one in 1980 to around three thousand to one in 1990–91, only to be made even more challenging after January 1, 1993 with the introduction of the New Mexican Peso decreed to be equivalent to one thousand former pesos. Through the years in this challenging environment businessmen in the border brick industry have been especially adept at conducting their transactions so as to maximize the difference in exchange rates between the two currencies. Indeed, one of them, Efraín Salinas, had so much success during several decades in the export brick business that he established a *casa de cambio* [currency exchange business] across the river in Hidalgo, Texas which is where my first meeting with him took place in 1992.

In an spiral notebook that was still in Omar Zapata's possession after having discarded most of his business records was an entry dated July 11, 1978 [dollar to peso rate twenty-three] itemizing estimated costs for producing one thousand fired brick at his own wood–firing *ladrillera* [not Tex–Mex]: molding—80 pesos [$3.50]; clay—70 pesos [$3.00]; water 10 pesos [$.43]; covering brick with plastic sheeting—10 pesos [$.43]; plastic sheeting—10 pesos [$.43];

carting of unfired brick to kiln—30 pesos [$1.30]; setting of brick into kiln—20 pesos [$.87]; sealing of kiln loading ports—40 pesos [$1.74]; firewood or leña 150 pesos [$6.52]; firing the kiln—50 pesos [$2.17]; unloading kiln—25 pesos [$1.09]; paint for coloring brick—25 pesos [$1.09]; dipping brick in paint solution—25 pesos [$1.09]. This amounted to a total of 704 pesos [$30.61]. Added to this total were the costs for workers' social security and other incidental expenses amounting to 96 pesos [$4.17]. This resulted in a grand total cost of 800 pesos [$34.78] per thousand brick produced. The common building brick Zapata was making and selling measured 8 x 2–1/4 x 3 inches in length, thickness and width. They were produced in a variety of natural and painted colors and sold for around $100 per thousand, thus confirming the high profit margins that prevailed at the time.

Conclusions

A good case can be made that despite its uneven record regarding plant unionization, the labor movement of the 1960s and 1970s did positively impact real wages of workers in the industry. Wage increases were the focus of every contract negotiation, every strike, and every worker petition to employers. In any case, real wages are remembered as having peaked in the years before the 1982 devaluation. The following statement in the mid–1990s by a fifty–seven year old piecerate brickmaker with thirty–three years of work experience at one of Reynosa's largest *ladrilleras* expressed this well:

> In those days [1960s–70s] I earned 400 pesos [$32.00] weekly. That was big money. Money was worth something. With one hundred pesos [eight dollars] you bought your groceries and still had money left over. Nowadays with 400 New Pesos [$100.00] that you earn you can't buy much because everything is so expensive. Nowadays it doesn't go anywhere, your budget is shot. With my weekly earnings here of 200 New Pesos [$60.00] I can't buy the groceries we need. The consumer is lost. Things are very uneven. You are forced to eat with the little that you earn.

After the collapse of the boom market, the ranks of the producers thinned out in the 1980s, including Omar Zapata among many others. The process of attrition resulted in a considerably downsized Mexican brick export industry coping with a normally fluctuating market that followed in broad outlines the ups and downs of the business cycle. This translated into alternating months and seasons of high sales and low sales but rarely periods of no sales. But the speculative building market of the 1970s that had latched on to Mexican handmade brick because it was relatively cheap and available was no longer operative by the early 1980s. The Texas market niche that remained for Mexican export producers was reduced to custom/architectural projects driven more by aesthetic demand than by availability and price. These conditions were, nevertheless, still compatible with the view of a persistent artisanal producer like Toño Reynoso's that "The brick business is not to sell them. The brick business is making brick and brick sell themselves."

Notes

1. This profile is of men currently employed by *ladrilleras* who either worked or are currently working as *hechureros*. A few of these men now work as watchmen or do light work in the *ladrillera* other than brickmaking, an occupation from which they have retired.

2. The CTM was founded in 1936 by Vicente Lombardo Toledano and Fidel Velásquez, and during the presidency of Lázaro Cárdenas became the largest and most politically influential organized labor organization in Mexico. In subsequent decades of the twentieth century this status continued in a formal corporate alliance with the ruling party, PRI or *Partido Revolucionario Institucional* (Barry, ed., 1992:183–187).

3. In addition to the Tueme plant in Reynosa, two other successfully unionized plants were Terra–Mex in Ciudad Miguel Alemán and *Ladrillerias Camargo* in Camargo.

4. It can be inferred from a comparison of the 1970 and 1974 contracts that there were still nonunion employees working at the *ladrillera* in 1970. The first clause of that contract in which the enterprise agrees to exclusively employ union workers contains the following qualifier: "*pero se reconoce preferencia a los trabajadores que laboran actualmente, independientemente de que sean or no, sindicalizados.*" That qualifier is absent from the 1974 contract.

5. When asked the question about their union status, either in the past or present, border brick workers respond in terms of a dichotomy: union vs. free [*sindicato* vs. *libre*]. What they mean by *free* is the independence to negotiate the terms of their employment directly with the employer. Most of them do not like the idea of belonging to a dues–charging bureaucracy like the CTM which was, in effect, the PRI–Gobierno. As artisanal pieceworkers they tend to view themselves as independent workers who earn according to what they produce pretty much on their own schedule. Most of the workers I interviewed who have worked in unionized and non–unionized *ladrilleras* preferred the latter experience. There is no question, however, that negative brick worker attitudes regarding unionization reflect the impact of anti–union sentiments of employers, as well as informal blacklisting of labor militants. Also, many workers were negatively impacted by the union–busting shutdown of several major *ladrilleras* in the Reynosa area during the 1960s, and tended to blame the union for the shutdowns.

6. In Mexicali, according to Wilson (2005:160), since the 1960s a minority of brickmakers belonged to CNOP [*Confederación Nacional de Organizaciones Populares* or National Confederation of Popular Organizations], a sort of middle class interest group within the PRI corporate system but in the 1970s joined CROC [*Confederación Revolucionaria de Obreros y Campesinos* or Revolutionary Confederation of Workers and Peasants] that was more attentive to the needs of small industrial producers [e.g., exempting them from paying social security for workers and allowing them to transport brick in their own trucks]. I do not know if either of these organizations were present among Reynosa's

brickmakers, although a similar type of organization called *UNE* was operative in the 1990s among the brickmakers of San Pedro Las Colonias [near Torreón] Coahuila.

Chapter Seven
After the Collapse: Attrition and Survival of the Wood–Firing *Ladrilleras*

For the uninitiated visitor to the brickmaking areas of Reynosa's periurban *ejidos*, the images and sounds impacting one's senses are paradoxical and sometimes shocking. Mesquite thickets [*chaparral*], scrubland [*monte*], and pasture [*agostadero*] give way to extensive areas of *ladrilleras*, many of them intact but abandoned and deteriorating, with others being used as trash dumps. Open areas of lightly populated, dispersed ranch houses are juxtaposed with densely populated areas of shanties or squatter settlements. These areas are connected through a network of unpaved roads and truck paths plied by varieties of vehicles, including bicycles and horse–drawn trash–hauling carts [*carretones*], and pedestrians. Cattle, horses, or goats graze in fenced fields or corrals sometimes near rows of idle kilns, and pig pens occupy abandoned clay pits. In short, these are semi–rural, mixed economic zones of competing low–tech industrial, agricultural, and service activities.

In my research the decade of the 1990s represents "the ethnographic present" or the period of most prolonged, continuous experience with brick plants in operation. Most of these were small–scale, wood–firing plants located in five semi–rural, periurban neighborhoods known both as *ejidos* and *colonias*: El Banco, Calabazas, Los Longoria, Rancho Grande, and Cavazos [Map 2]. A majority of the residents in these areas had work experience in the brick industry even though, for the previous decade, employment opportunities in brick plants have been cyclical and never open–ended as during the boom days of the 1960s and 1970s. Those plants that survived the 1980s, like those of Tano Arriaga and Toño Reynoso, remained viable either by self–exploitation and depending on financing from clients, as in the case of Arriaga, or by operating at a reduced level of output except when upswings in the market supported limited expansion as in the case of Reynoso. These low levels of output typically revolved around production of adobones for the domestic construction market, supplemented by production of other brick for export. Late in the decade of the 1990s there were signs of renewed export demand some of which favorably impacted the few remaining small producers but was mostly absorbed by *Ladrillera Reynosa*, then the largest of only two remaining gas–firing plants from the boom period.[1]

The decade of the 1980s was marked by a steady decline in the export market, caused by a convergence of several factors including the fallout from a massive multimedia campaign against the importation of Mexican handmade brick waged by the Brick Institute of Texas on behalf of the mechanized brick industry, the 1982 drop in oil prices and the Mexican peso devaluation, and a depression in the Texas economy and building industry (see Cook 1998:Ch.12). Omar Zapata in discussing the collapse [*desplome*] of the brick market observed: "Business started downwards, downwards. Houston stopped buying. My company [Tex–Mex] built around 100 apartment buildings, large units [in the late 1970s and early 1980s]. We went to see them around 1982–1983 and eighty of them were vacant and twenty of them were being offered at half price. Everything collapsed. People left Houston, a lot of people, and business collapsed." So, by the mid–1980s when Zapata retired from the industry, the *fiebre* [fever] of previous decades had turned into the *desplome*.

From the perspective of satellite imaging, the Los Longoria–Calabazas landform created by the meandering Rio Grande/Rio Bravo, appears as a misshapen peninsula (see Maps 2 and 5). The landform measures 1.95 miles across its mid–section between its western and eastern river boundaries, and .81 miles from the river on the north to its southern boundary with other municipal entities. Although physically connected to Reynosa City, this area has been protected from urbanization by its mix of *ejido* and private ranching properties. It was in Los Longoria–Calabazas that the most dramatic growth of wood–fired brickmaking occurred after 1950, and where the satellite field or *campero* system saw its most intensive development. This remarkable landscape, that I first visited in 1987, encompasses a plethora of abandoned *ladrilleras*, many turned into trash dumps, dispersed clusters of ranch buildings and residences, several operating *ladrilleras*, and densely inhabited shantytowns on its southern fringe. Its westernmost section was historically dominated by the *Hacienda Longoria*, and the heirs of this, by then abandoned, cattle–raising and fodder–growing enterprise were absentee landowners.

Los Longoria–Calabazas can be reached from the international bridge by driving west along the levee road, then onto the levee [*bordo*] passing just north of Reynosa's red–light district to a point where the levee veers south, and continuing west on an unpaved road until entering the area of the *ladrilleras*. The distance traversed on this trip is about one and a half miles. My project's field map from 1993 [updated 1995] showed fifteen separate brick plants and three plant clusters, one of which had ten kilns, within the Los Longoria–Calabazas brickmaking area. At that time nearly half had been shutdown for periods ranging from two to twelve years, and two of the clusters had temporarily suspended production pending the outcome of a lawsuit pitting several *ladrillero* renters against the landlord's widow. There were eight plants in operation within the area during 1993–95, just a fraction of the number operating in the 1970s.[2]

Changes in Periurban *Ladrilleras* and *Ladrillero* Livelihood

Many of the *colonias* that make up the city of Reynosa were establislished during the first–half of the twentieth century through the expropriation of private or vacant [*baldio*] lands and the establishment of *ejidos* or agrarian reform communities. Many *colonias*, like Rancho Grande, Los Longoria and El Banco are comprised of a combination of *ejido* and private land [*pequeñas propiedades*]. During the 1990s land tenure reform was undertaken through a government program of *regularización de tenencía*—a process of negotiation involving the *Comisión Agraria* of the municipal government, the *Comisariado Ejidal* representing the ejidos, and the *Procuraduría Agraria* of the federal government, as well as political parties and other interest groups. Since then the process of 'regularization' of tenure has proceeded unevenly, with mixed results, to convert *ejido* tenure into private possession or de facto private ownership.

Los Longoria-Calabazas

Curiously, only one document in Los Longoria's Expediente Agrario directly mentions the brick industry, a 1993 petition by seventeen non–*ejidatario* residents who were brickmakers by trade [*avecindados de oficio dedicados a la fábrica de ladrillo*] soliciting help to become owners of the property that they had possessed for thirty–five years. This document is an outgrowth of late twentieth century revisions in Mexico's Federal Agrarian Code [*Código Agrario*] encouraging the regularization and privatization of tenure, and is especially congenial to the interests of settlers [*avecindados*] who seek credentialed possession of residence lots [*solares*]. According to Ing. Patricio Garza, director of Reynosa's Agrarian Reform Office [*Delegado de Reforma Agraria*] (1993 interview with author):

> The brick plants are located on *ejido* land. If these brickmakers worked out agreements with the ejidos for the exploitation and use of land for brickmaking, or if they did not arrange this, and if they have more than five years of possession without problems with their neighbors or *ejidatarios*, the law automatically protects them as possessors of the land. At a determined time, the *Procuraduria Agraria* will issue confirming documents to them, this is one of the new examples provided in the law to make them legal possessors.

A controversial case in point at the time involved a deceased landholder, Domingo Morales, who had rented his land over the years to various *ladrilleros* who had moved into the area from elsewhere. Morales' widow was the principal litigant contesting the ladrilleros' petition to become *posesionarios* of their rented *ladrilleras*.

A sample of sixteen brick industry workers in the 1990s, employed at five different plants located either in *Colonia Rancho Grande* or *Ejido Los Longoria–Calabazas* shows considerable diversity in background, except for the fact that none were natives of Reynosa. One worker was born in the state of San Luís

Potosí but was raised in Reynosa from the age of three months in Los Longoria–Calabazas where he now works; and two others were born and raised in communities close to Reynosa in the state of Tamaulipas. Four were from Coahuila, underlying that state's historic importance as a source of labor for the Reynosa industry; and two were from Nuevo León. The remainder were from Chihuahua, Durango, Michoacán, Aguascalientes, and Zacatecas.

In terms of age and work histories the data for these sixteen workers, only one of whom was a woman, is equally diverse. Two of the workers interviewed, Manuel Varela and Hermilo Ramos Castañeda, had worked for *Ladrillera Reynosa* since the early 1960s; they were the exception to the rule for brickmakers who typically have much shorter periods of employment at one particular plant. Indeed, the typical brickmaker tends to circulate around several brick plants, and between employment in other sectors like agriculture and construction, during his career.

The case of Mario Armendáriz who was thirty-seven years old when I interviewed him at *Ladrillera Cantú* in August, 1993 is representative. He came as a child to Reynosa and *Ejido Los Longoria*, and started working in the brick industry at the age of twelve. He worked with Juan Tueme for five years, then in several ladrilleras on the land of Domingo Morales, and with Francisco Macías also in Los Longoria. He had been working with Cantú for only two months at the time of the interview. He had six children to support and said that he could more or less maintain his family through his earnings as a brickmaker, which he prefers to construction work. In his words: "You earn more here than in construction. Here, there is no one bossing me around. You work alone. I can cover up my clay and leave without anyone saying anything." Unlike many brickmakers who did not work on Mondays, Mario worked six days a week and, if the weather cooperated, made eight hundred brick daily. At the going pay rate of the peso equivalent of 35 dollars per thousand, his weekly earnings were as high as 165 dollars.

Agustín Nuñez, born in 1961, was foreman at the Cantú plant where Armendáriz worked, and came to Los Longoria with his parents and six brothers from Guanajuato when he was eight years old. He worked for six years in brickmaking and cutting mesquite for firewood [*leña*], with his father until the latter's death. Then he worked for three and a half years at the Reynoso plant and two years in construction before going to work for Cantú for the next fifteen years. When Cantú closed his plant in 1990 for three years, Agustín worked as a truck driver. Cantú reopened the plant in early 1993. He leased it on an annual contract basis from Sebastian Reynoso and paid rent in fired brick. At the time I visited this plant in August, 1993 most of the brick were being made in a nearby brickfield in Rancho Grande, also leased by Cantú, where five or six brickmakers were employed on an irregular basis. They dug their own clay with hoes [*asadones*] and shovels [*palas*] and each prepared his own clay batch. This was strictly a low–tech operation. The unfired brick were manually loaded on a truck and taken to the main plant, where they were wood–fired. Five truckloads of mesquite were required, at a cost of 250 to 330 pesos [$75 to $100] for each firing of a loaded kiln. Most of the brick [eight inch or nine inch) were sold to a trucker–intermediary from Corpus Christi, Texas but adobones were also sold to local building materials suppliers [*madererías*]. The best selling product was the nine–inch *barra.* There were sufficient brick in inventory for fifteen to twenty trailer loads [twelve thousand brick each] at the time of my visit but, according to Agustín, only three had been sold since the plant went back into production. The going market price per thousand brick at the time was in the neighborhood of 150 pesos [$45].

Nicolás Rodríguez, born in 1923 in Morelia, Michoacan, began to work as a *ladrillero* in Reynosa and Hidalgo, Texas in 1944, after working for two years as a *bracero* in the U.S. Midwest. His parents migrated to Texas in the early 1940s and settled in the Rio Grande Valley. From 1944 to 1948, according to Nicolás:

> I worked in the putting–out system, I worked as a putter–out [*Yo trabajé a la maquila. Yo andaba maquiliando*]. I began to make large *adobones* to teach myself how and then they changed me to make brick twelve-inches long. Those were for export to the other side. We exported them crude to Hidalgo. We made them over here and transported them unfired to Hidalgo for firing in a kiln. Ramiro Rodríguez transported them for firing in Hidalgo.

This is a description of one of the earliest examples of the *maquiladora*–like arrangement in brickmaking that later in the same decade would involve Sebastian Reynoso and other repatriated *ladrilleros*. This preceded the establishment of the Border Industrialization Program in 1965 that officially established the twin–plant or *maquiladora* system as the backbone of the new border economy [cf. chapter 5].

Nicolás became a specialist in stoking/firing kilns [*quemador*] but also supervised the plant. He explained his arrangement with Ramiro Rodríguez as follows:

> I made brick and he fired them. But since I am also a stoker, he paid me to do it. Since Ramiro was the owner of the brick plant, he put me in charge of setting and firing. He'd show up and ask me: 'How many do you see?' I'd tell him how many. We loaded the trailer with unfired brick on this side of the river and I would go with him in the trailer to unload them, set them in the kiln, and fire them.

In Hidalgo, Ramiro Rodríguez had three kilns each with a capacity of around seventy thousand brick. Typically one kiln was fired monthly. Since Rodríguez was a trucker he transported the brick directly to clients in Corpus and Houston.

By the mid–1960s, Nicolás was working as foreman and stoker with Ramiro Rodríguez in Reynosa where he rented kilns. According to Nicolás: "Ramiro would give me money to pay workers. That is, the work that the molders did was paid for by him through me. I was still not working as a brick molder." But in the late 1960s, according to Nicolás, Ramiro "no longer wanted us to make brick for him . . . He became involved in buying brick from other producers, from the *camperos*."

In the late 1960s, then, Nicolás started his own *ladrillera* on land rented from Domingo Morales. He worked for the next fifteen years or so, with his five sons [he also has two daughters], and produced *adobón*, mostly for sale in Mexico. According to Nicolás, his clients, mainly masons [*albañiles*], would simply come to the plant and ask: "Do you have material? Put it on my building site." Then, he explained, "I hauled it to the building site myself. I hired a truck to transport it to the building site." Nicolás' brick plant had a clay pit and a small kiln with a capacity of 14,000 adobones. He had to purchase mesquite for firing the kiln. He paid Domingo Morales in cash for each kiln load; the amount paid was 10 percent of the material in the kiln [*diez porciento del valor del material en el cocedor*]. Morales would monitor brickmaking activity every two weeks or so. According to Nicolás, "He [Morales] knew when we were going to fire the kiln; he came every two weeks or every month. He appeared and asked how we were doing."

Following the death of Domingo Morales in the early 1990s his widow decided to no longer rent land to make brick. As mentioned above, the *ladrilleros* protested and petitioned the *Procuraduria Agraria* for possession of the plants. Because of the dispute some plants were closed in 1992 but by then several plants in the vicinity of Morales' land had already been converted into trash dumps.

Most of the *ladrilleros*, like Nicolás, who worked on Morales' land, were also squatters who lived next to the *ladrilleras*. In the summer of 1993 many of these *ladrilleros* had become day laborers since they were unable to work in the *ladrilleras* due to the lawsuit with Morales' widow who no longer wanted to rent to them. During a street corner discussion with these men, who were somewhat reticent to talk to me for fear that I was representing either the government or Morales' widow, I asked the question as to whether the land was private or *ejido*. One of them responded as follows: "Here, at present, we are associated. For a year and a half we have had a committee advised by a lawyer. The owner [Domingo Morales] died. Then, the widow was going to bring heavy equipment to bulldoze the houses and the kilns. That was stopped thanks to us. They wanted to run us out of here." Another member of the group expressed his version of the situation: "We are not working since the land is in dispute. Given the number of years that we have lived here, we believe that we have rights 'as if' we were owners [*como dueños*]. There are twenty of us, seventeen are brickmakers who live in this area. As possessors [*posesionarios*] we have rights. We do manage to work some but there is no demand, there are no sales."

I interviewed Nicolás Rodríguez later that same day in July 1993 at a plant rented by his son–in–law, Francisco "Pancho" Macías, from a Los Longoria *ejidatario*. This plant was located about one kilometer away from the disputed Morales property. Unlike the arrangement Nicolás and other members of the street corner group had with Morales, which permitted the renter to dig for clay on the land rented, Macías rented only the space and the kiln with a capacity of ten– to twelve–thousand brick. Clay could not be dug there but had to be trucked in from another source. Rather than mesquite, scrap wood—usually discarded pallets purchased in Reynosa's maquiladora zone—was used to fire the kilns. Nicolás, who was in charge of firing, pointed to workers making brick, and said: "Among those three they make 1,500 brick daily; monthly I can fire two kiln loads."

Macías, born in 1955 in Aguascalientes, started working in Nicolás' plant at the age of eleven but by the time he married Nicolas' daughter at the age of eighteen he was working at *Ladrillera Reynosa*. In his words: "I left to work on my own because I wanted to have children and I had to assure that I could support them in terms of food and education. With Nicolás I earned only for my elders. There was a larger export brick plant [i.e., *Ladrillera Reynosa*] where they offered social security for illnesses. Once that I married I sought insurance coverage." Francisco worked at *Ladrillera Reynosa* for two years as a molder [*hechurero*], then served in the Mexican army for a year. After completing his military service, he went to Dallas, Texas with his brother to work in construction, returning to Reynosa in 1978 with a used truck to work as a construction sub–contractor.

In 1983 Macías went back to Dallas, Texas where he worked as a janitor for five years straight without returning to Mexico. During that period he acquired permanent residency status in the U.S. In 1989 he returned to Reynosa to work again as an *hechurero* [*Desde chico me gustó este trabajo*] but this time on land rented from Domingo Morales and with two employees. At the time I interviewed him in the early 1990s he had been working on his own account for three years, sometimes with three or four employees.

Macías's plan was to produce brick for the export market but he recognized that he needed contacts on the U.S. side of the border to do so successfully. His clients in the early 1990s were all local—either contractors or masons, or lumber yards [*madererías*]. Most of his clients bought brick directly from him at his plant site. His goal was to fire ten thousand to twelve thousand *adobones* monthly. He was able to achieve this so long as the weather cooperated. During periods of heavy rainfall, production of new brick ceased and he had to rely on inventory already set in the kiln ready to fire. For example, in the month pervious to my interview, rains had forced such a shutdown but he was fortunate to have a nearly full kiln that helped him continue paying his workers after weather conditions improved. Despite periodic rain interruptions he estimated that his average monthly output was eight thousand to ten thousand *adobones*. The two workers he employed, worked three or four days weekly. One of them produced five hundred pieces daily and the other produced a thousand pieces. Macías set aside *adobones* from each firing for the landowner. He pointed to a stack of three thousand brick and commented that they represented the owner's share from three kiln firings.

Carlos Aguilera García, born in 1944 in Durango, started working as a *ladrillero* at the age of fifteen in Guadalajara but came to Matamoros in 1965 where he also worked as a trash hauler [*carretonero*] and as a sorter or recycler [*pepenador*] of scrap metal. He moved to Reynosa in 1980 and continued the same job pattern: *ladrillero*, *carretonero*, and *pepenador*. At the time I interviewed him in 1993, he had been working for two years in his cousin's small brick plant mainly making *adobón*. Previously, he worked for Anita Soto. He lived nearby with his wife and five children. One of his sons worked with him at the brick plant, but the rest of his children still attended school. According to Carlos, the advantage of making brick for a living was that he was able to work at his own pace without anyone telling him what to do. His daily output of *adobones* was three hundred, whereas in his younger days he made five hundred or more.

Regarding his previous three–year stint in Los Longoria as a trash hauler and sorter, Carlos explained: "When there was no work available in the brick plants I went to work in a trash dump. I earned a little money but it was tough. On the other hand, if you do nothing you don't have anything to eat." He also tried to raise pigs to supplement his income but was unhappy with the results.

> I was working at a brick plant and I bought several little pigs to raise until they were large. I bought them feed for the week but they finished it in less than a week. So, I had to get rid of them because I decided that I was working simply to feed the pigs. A neighbor lady advised me to stop buying them feed and, instead, go to the trash dump, gather up garbage, and feed it to the pigs. That was the solution to feeding the pigs.

Carlos saw no alternative to supplementing work in brick making with pig raising or other odd jobs to earn cash. His kiln was too small, and sales too slow and unrewarding, to provide him with anything other than a survival income. As he put it: "With a kiln like this, that you fill with a little kindling, it's hardly worth the effort because you get paid less than you want. This makes life difficult. In this brick plant I take care of myself as best I can. I come here to leave things not to take them away. Where is the business? This way it's difficult!" [*Yo aquí me defiendo como puedo. Aquí viene uno a dejar, no viene a llevar. ¿Dónde está el negocio? Así está difícil.*]

El Banco

The situation in 1993 was equally bleak in Colonia El Banco where the landscape, even more so than in Los Longoria, epitomized deindustrialization and a process of transformation into a periurban dumping ground or wasteland. A plethora of large–capacity kilns that once billowed smoke from the wood–firing of millions of brick, mostly unused and deteriorating since the 1980s, are surrounded by vast junkyards and the deep clay pits that provided the raw material for the once–thriving brick industry are gigantic open–air receptacles for trash and garbage hauled in horse–drawn carts from the nearby city. Ironically, the smoke that hangs in the air these days comes mainly from slow–burning fires in the trash dumps, not from kilns that are firing brick. It is, indeed, a sad spectacle best captured in the almost shouted words of a young woman in a group of residents who gathered around me during my first visit to the area, thinking perhaps that I was some sort of change agent: "Sir, look around you and you will see that we are truly the 'forgotten ones of Mexico'" [*Somos los verdaderos olvidados de México*], alluding to Buñuel's gut–wrenching classic movie, *Los Olvidados*, about the trash pickers of Mexico City. I subsequently found out that the young woman who had made this dramatic statement was Clara Salazar, a member of the Salazar family of brickmakers and storeowners who, in fact, were managing somehow to make a living in this depressed and depressing environment. [See Appendix 1. Photos 15–17 provide images of this deindustrialized wasteland that still supports an industrious, if marginalized, population].

A majority of El Banco's current residents were recruited to work in brick plants that sprouted up during the 1970s boom. They were provided with housing by entrepreneurs who rented land from *ejidatarios* to make brick typically by paying rent, either in cash or kind, equivalent to ten or fifteen percent of each thousand brick manufactured. After the boom ended in the early 1980s, the *ladrilleros* who had become residents of El Banco stayed on to try to make a living by making brick, also by paying rent to the *ejidatarios*. Many others commuted to the *ladrilleras* there daily and lived in other *colonias*. Their brick businesses depended on *ayuda* [assistance in the form of cash advances] from truckers or other intermediaries, many of whom lived in the Rio Grande Valley. Typically, these surviving *ladrilleros* molded and fired a combination of *ladrillos* and *adobones*, the latter being most popular then in the greater Reynosa market. They supplemented brickmaking with other odd jobs in agriculture, construction, recycling trash, selling soft drinks and other foods and sundries in small stores, and so on.

Matías Rivera is a native of Pánuco, Veracruz where his father was a peasant agriculturist and brickmaker. At the age of thirteen after finishing elementary school, he started to work in the *ladrillera* where his father was employed. The *ladrillera* used *chapapote* or crushed bamboo–like reed [*carrizo*] as a fuel for firing brick called *tabicones* which are somewhat smaller than *adobones* in

Reynosa. Another difference from Reynosa's wood–firing plants was the use of draught animals to mix clay. Five men worked as a team [*cuadrilla*]. Two were *paleros* [shovelers] who helped to prepare and move clay to the molding area, two were *tendedores* who molded and laid out brick on the drying floor, and one was a *cortador* [molder or 'cutter']. The team of five produced 5,000 brick daily. Matías described the process as follows: "There was a large pile and the earth was moved with wheelbarrows. You had to push the earth up the mound with wheelbarrows; that was the hardest part of the job. On the large pile was where the wheelbarrows dumped the clay. And the draught animal [*bestia*] mixed it. The *tendedores* and the *cortador* ran and ran. Everything was very fast."

Since coming to Reynosa in 1977, Rivera worked alone on rented land in El Banco. The land he rented had an old, crumbling kiln that had been converted into a pig pen. He rebuilt this kiln that has a capacity of twelve thousand pieces, and subsequently built another kiln with a capacity of twenty thousand pieces. I interviewed Matias on two occasions in July, 1993 while he worked. His plant had a clay pit about fifteen feet deep where he dug out his clay supply with a hoe and shovel, and mixed his clay with water drawn from a nearby well. Although doing the entire process of clay digging and preparation himself, Rivera still managed to mold twelve hundred brick per day. He also performed every task in the drying, setting, firing, and drawing of brick in an unusual one–man operation. He had one son, who was employed in a *maquiladora*, and who he did not pressure to work with him at the brick plant. [See Appendix 1. Photo 18 shows Rivera digging clay exactly at the spot where he mixes it, and photo 19 presents a view of his unique one–man plant].

Rivera's rent payment varied from a rate of 15 percent per thousand to ten percent of each kiln load. The first time I interviewed him he explained the rent payment as follows: "When you have pieces, the rent is 15 percent of the material you made. Of every one hundred brick made, fifteen are set aside for rent." But, during a subsequent interview in response to a question about how many brick from each kiln load he sets aside as rent, he pointed to his kilns and responded: "Well, for example, in this kiln I handed over 900 of the 9,000 adobones fired. In this other kiln, I fired 20,000 and handed over 2,000 to the owner. That is 10 percent." Despite this discrepancy, Rivera did not consider rent to be expensive. His most punishing cost is the firewood he uses to fire his kilns ["*La leña es lo que castiga uno. Me sale muy caro. Está muy duro.*"] At the time, he was paying thirteen thousand brick for two truckloads of firewood; that amounted to almost one–half of the output of his two kilns.

Rivera did work as a supplier, for a time, with the trucker and dealer from Mission, Texas, Ramiro Rodríguez but refused to make brick for him or others with their particular frog [trademark indentation or *placa*]. His brick is made with either with the Mexico export mark or with no mark at all for the domestic market. This practice, he felt, gave him more freedom to sell his brick to any buyer that showed up, and assured him that all the brick he made was his to sell.

He had a client from Matamoros who was regularly buying for government drainage contracts—but this had ended at the time he was interviewed and his realistic assessment was, "When you have no clients, nothing moves." Most of his clients are from the Reynosa area, and typically buy only a thousand or fewer brick for small construction projects.

Hilario Salazar is one of sixteen children of Joaquín Salazar who was born in Romita, Guanajuato in 1913 and arrived in Reynosa in 1920. The family lived in various *colonias* of the city before acquiring rights to property from *ejidatarios* in El Banco. Joaquín learned brickmaking in El Banco with a *ladrillero* named Lino before Hilario was born. Hilario learned brickmaking in his father's plant along with his siblings, including his oldest sister Clara who also worked in the plant until she married. In 1993, the Salazar plant had two locations: one in a residence lot and another about two blocks away; both were private property. The plant had a wood–firing kiln, and the Salazars had an arrangement that allowed them to clear land for firewood, mainly mesquite. They cut one or two truckloads every three or four months, doing so when the weather was not conducive to brickmaking. They accumulated firewood in anticipation of a kiln firing every three or four months. In previous years, when five or six family members were working, the kiln was loaded with fifty thousand to sixty thousand brick every four to six weeks. In 1993, with only Hilario and two of his brothers working, they divided the kiln between thirty thousand to forty thousand nine–inch brick and larger *adobones*. They sold brick for forty–five dollars per thousand to their best clients who paid in advance and bought large quantities. Locally, however, they sold *adobones* for fifteen centavos each [about five cents].

Hilario's description of how this *ladillera* operated then and now applies to most others in El Banco: "People from the other side bought material from us. They made it possible for us to work. That was how we used to work. Now, it's different. We work on our own account. A client shows up, asks for so many pieces of material, and buys. Since I am working in a very measured way it is a little bit difficult." In 1993 the Salazars were buying clay rather than digging it on their own land. On the day that I interviewed him, Hilario was molding nine–inch brick with one of his brothers [see Appendix 1. Photo 20].

In 1993 Ascensión Navarro, whose earlier career was discussed in chapter 5, had still not sought the status of *posesionario* and was still a renter [*arrendatario*], occasionally paying a peso equivalent of fifteen dollars to the owner. He only occasionally fired one of his three kilns, all of which were in poor condition, using discarded pallets and other wood extracted from the nearby trash dumps that he hauled to his plant in an old, beat–up pick–up truck. When I asked him why the brick business was so unrewarding now compared to past years, he responded that the buyers that used to buy no longer did so and that the ones that remained no longer paid high prices. Moreover, today's intermediaries only bought now and then, and were unable to sell at high prices on the other side. He also added that competition from cement block had driven down the price of clay brick. All of these conditions were, of course, beyond the control of the brickmakers themselves.

Navarro's business consisted of occasional sales of small lots of brick, one or two thousand per transaction. He sold for $35.00 to $50.00 per thousand but typically had to load, deliver, and unload the brick for that price. Given the poor condition of his kilns, Navarro's output was limited. He no longer bought fuel in the form of cut mesquite but simply scavenged in the nearby trash dumps for discarded pallets or other pieces of wood to fire his kiln. He had resigned himself to trying to keep a small inventory of brick to supply occasional buyers. Nevertheless, even under conditions where earning a living was precarious, Ascensión emphasized the advantage of brickmaking, appreciated by most workers who remain in the industry: "In this work nobody scolds you. One works how he wants to, and when he wants to. The days that he doesn't work there is no one who asks him 'why'."

José Velásquez Bauzá, whose early career as a piecerate brick molder was also covered in chapter 5, by the mid–1980s was operating a plant on his residence lot for which he had purchased rights of possession from an *ejidatario*. He extracted and hauled clay from another parcel of land owned by the same *ejidatario* for a 10 percent share of his output of fired brick. José's wife and two eldest sons worked with him part–time making *adobones* for the local market that he sold directly for buyers for their own use or, at times, to resellers who contracted for large lots.

When I met and interviewed him in 1995, his last contract with a reseller had ended three years earlier. In Velásquez's words: "I started to work with a guy from Rio Bravo who helped me make brick. He paid me like he was the foreman. I molded the brick for him, I set green brick into the kiln, I fired the kiln. I did everything and he paid me for my work and he hauled away all of the brick. He buys and sells different kinds of building materials, including sand." Velásquez finally ended his business with this individual because he was paying only 180 New Pesos [$23.44] per thousand pieces, a price that essentially left him without any profit. In any case, during the three–year period that he worked with this reseller, Velásquez estimated that only twenty thousand or so *adobones* were sold to him.

In 1994 Velásquez stopped making *adobones* himself due to a back problem but maintained a low volume of output with the help of an older worker who made 200 or 300 *adobones* on his own schedule for which Velásquez paid him 90 New Pesos [$11.72] per thousand. His selling price at the time was 350–380 New Pesos [$45.57–$49.48] per thousand. Velásquez emphasized that the brick business was very slow, and provided only a small portion of his household income—supplementing contributions from his employed elder children, from his wife's trash recycling, and from the soft drinks and a few other items he sold out of his house, mostly to the trash haulers who regularly passed by on their way to and from a nearby trash dump. He noted that it was only about a year and a half or so ago that one of the *ejidatarios* decided to start a trash dump in an abandoned clay pit on his land and that, afterwards, others started doing the same.

Velásquez summarized his overall situation in the mid-1990s as follows:

> The brick business is very slow. It just helps us out a little. It is not really a business because I do not sell more than a thousand brick every three or four months. Buyers show up and want to buy fifty, 100 or 200 brick. Do you think that is sufficient to maintain one? That is why I spend more time here in the house than I do at the brick plant. My wife recycles things from the trash dump. She picks up discarded motors that I rebuild, as well as aluminum and copper. That helps us out more than the brick plant. And then one must go to pick up ice for the soft drink cooler and prepare lunch for the kids. So, I spend relatively little time with the brick business.

Although the brick business was slow in 1995, it had been good enough during the 1980s to enable him to save enough to make down payments, and eventually payoff on installments, a used car and a used truck that he still owned and operated. He also acknowledged that molding brick had helped him to put two of his sons through secondary school, and was quite proud that this had led them to get salaried jobs—one in the *maquiladora* sector and the other as a printer. His goal was to provide them with job opportunities outside of brickmaking because the work was too hard and sales too limited to provide a steady income. [See Appendix 1. Photo 21 shows Velásquez Bauzá at his brick plant].

Tano Arriaga and His Cavazos Plant

Of the many *ladrilleros* I interviewed in the 1990s in Reynosa's periurban colonias, Venustiano 'Tano' Arriaga of Los Cavazos stands out for having transformed years of hard physical toil, first as an employee and later as worker–owner in a plant with both family workers and employees, into a decent living for his wife and seven children. His comfortable two story house on the main street of Cavazos, that he constructed himself, provides material evidence of this, as does his role as a baseball *aficionado* who sponsors and manages a local amateur league team. Tano's success in the brick business is even more remarkable in view of his lack of formal schooling and his inability to read, write, or keep accounts, and the fact that his brick plants have always been on rented land. So, his success as a brick producer is a tribute to a combination of factors including his strong physique, native intelligence, capacity for work, and desire to succeed in maintaining his family and building his business.

Tano was born in 1940 in a ranching community in the state of San Luís Potosí where the main occupations were agriculture and goat herding. He was one of thirteen children in his family and, according to Tano, they were so poor that he was sent to live with another family to help care for their animals to earn twenty centavos monthly and his meals. When he was about ten years old, his family moved to the town of Pesquería, Nuevo León near Monterrey and found work in the handmade brick industry there. The brick plants had housing for workers where his family lived. He explained how he learned to make brick in Pesquería:

> When I started, it was nothing more than watching, I watched and listened to what the foreman was telling me. What interested me most was how he was making brick, all of his movements. I was interested and I knew that it was one thing to talk about making brick but very different to actually make them. The first day I made some without knowing how, I made a thousand from seven in the morning until two in the afternoon. I can't say whether or not they were well made. The maximum learning time is three or four days. If you don't learn in that time you will never learn. One learns if he really wants to work.

Over a period of four years in Pesquería from 1951 to 1955, the family lived and worked in two different *ladrilleras*. Afterwards, they moved to Miguel Alemán where Tano also worked successively in two different *ladrilleras*, met his wife, and started his own family at the age of twenty. One of these plants, *Ladrillera Falcón*, located on the highway to Camargo in *barrio* Los Angeles, had more than one hundred workers and their families, many of whom lived in rent–free row houses provided by the plant. At this point in his career Tano was a production leader among molders averaging twelve thousand to thirteen thousand brick weekly which was almost double the average output of one thousand brick daily.

In the 1960s, Tano was caught in a labor–management conflict at *Ladrillera Falcón*—this was the most intense period of unionization efforts in the industry. In his words: "I didn't like having problems with anyone, I was interested in working . . . they offered to pay me a higher salary so that I would inform them who among the workers was causing problems . . . I didn't want to do that. I didn't like to get involved in disputes."

Tano never joined the brickmaker's union but was, in his own way, a labor organizer. This is clear from his answer to my question regarding union activity in the Miguel Alemán *ladrilleras*:

> In those days there were people who were involved in union organizing. If that was acceptable to you and if the plant treated you badly regarding salary and all that well the union was an alternative. But those who didn't want to join, didn't. I didn't want to get involved in the union. I always said 'no.' When I had problems I resolved them on my own. To get a higher salary I went to talk to the owner. I would talk to him based upon the list of workers that were with me. I didn't know how to read or write, I didn't know anything, but I knew that fifty workers were with me and another forty or fifty were not. So when I spoke to the owner, he did not believe me, he didn't think that alone I could do anything. Then I showed him the list of workers that were with me, that supported me, and we got the raise. Afterwards, some workers could not stand others who

> were supporting the boss. I was in favor of being united but I never belonged to the union.

At that time, the brick industry was booming throughout the riverine zone and Tano saw an opportunity to move to Colonia Cavazos where *Ladrillera Industrializada Rio Bravo*, owned by Juan José Varela, had a workforce approaching six hundred workers, larger than *Ladrillera Falcón*, according to Tano's recollection. The plant had row housing for workers with families but also had buses to transport commuter workers from the nearby city of Reynosa. Tano was able to purchase a lot [parcela] in Cavazos outside the precinct of the plant but within walking distance and, with some help from the ladrillera owner, build a house for his family.

A combination of fluctuations in the brick export market and labor relations problems in the early 1970s at *Ladrillera Industrializada Rio Bravo* led to yet another transition in Tano's career involving a stint in Houston, Texas in the brickyard of a large importer, and by 1972 the start–up of his own brick plant back in Cavazos. He recounted the problems at Varela's plant as follows:

> I made brick and I told him that the salary was too low. I got together with other workers and we asked for a raise. He reacted badly and didn't give use a raise. So we got mad at him. But the time came when there was no business and he told us to find work elsewhere for a month or two and then come back and he would help us. But there was no work and we bought land from Varela and he built houses for us that we had to buy. Some of us paid them, others had not when Varela came and told us to look for work elsewhere and vacate the houses. I told him at that time: I paid you for this house and I don't want you around here. He got mad and pulled out a pistol. But he didn't pull the trigger. Then he told me that if and when he had work, he would hire me.

In 1972 Tano rented a vacant plant site which had a kiln in need of repair from a Cavazos *ejidatario*, repaired the kiln, and bought enough firewood and clay to make and fire eighty thousand brick. Tano explains his thinking about becoming an independent plant operator: "I worked in other plants. I learned the environment, how it was, how to work. Then I thought to myself that I could make brick on my own account and others would pay for them and order. I can make brick and I can fire them. And I did it."

Regarding his first independent venture he admitted that "I wasn't sure how the work would come out, I didn't know how long to fire brick. I didn't know any of that." Nevertheless, his risk–taking was rewarded: "The work turned out more or less well and I was able to get seven thousand dollars for my first kiln load. So I hired some workers." Among his employees were many of his friends from *Ladrillera Industrializada Rio Bravo*, some of whom were helping with heavy equipment belonging to that business and/or working for Tano on the side. In any case, for the following ten years or so, until the mid–1980s when "everything came down" [*vino todo abajo*], he estimates his production was sixty thousand to seventy thousand brick weekly or two hundred thousand to three hundred thousand monthly. He was justifiably proud of his effort: "I was working on a small scale but just like a larger plant. There were times that I sold more brick than they did because I had a lot of contacts in Houston, several places where I delivered."

During the peak years Tano operated as many as twenty kilns and employed as many as forty to fifty *ladrilleros* at his plant, and in addition had twenty camperos. His high output was a function of demand. His clients included several of the large gas–firing plants in Reynosa, as well as importers from Houston and elsewhere in Texas. They placed orders for one million to two million brick, but had to visit Tano's plant and leave a down payment to assure delivery.

In order to expand his output, Tano bought brick from other small plants in neighboring *colonias*:

> I went to Rancho Grande or to any other colonia and I asked the brickmakers if they wanted to make material. They would ask me how much I would pay. I told them I would pay a few pesos more per thousand than they were getting from other clients. They agreed and told me that their policy was to sell to clients who offered the best price. That is how I got twenty *camperos*. If they had trucks I paid delivery costs. If not then I sent my truck and they loaded it.

According to Tano, "There was so much demand for four or five years that I worked hard to supply it. I believed that the work was not going to end."

Tano described the campero system from the perspective of the business cycle as follows [cf. chapter 5]:

> During the boom everyone had camperos. There were camperos that looked for clients to make contracts with. Each person did the work with their entire family to sell to whoever they could. There were all kinds. There were big camperos that looked for clients as if they were owners, and they took charge of selling, collecting from those who sold, and pay the workers. There were times when the camperos had their own kilns on site. That was during the boom years. But after sales dropped, everyone was looking for buyers. It was no longer a business. Since then it has only been a question of sustenance.

This statement is, in effect, a summary of Tano's own experience as a *ladrillera* operator.

After 1985 Tano's business slowed down considerably to the extent that he depended entirely on his own and his sons' labor supplemented occasionally with two or three hired workers. For a time, he entered into a relationship with a trucker from Mission, Texas, Ramiro Rodríguez but this ended in the early 1990s when Rodríguez died of a heart attack. At the time of my interviews with him in 1993, Tano was operating his plant not so much as a business but simply to sustain his household. This is how he summed up the situation at that time: "I sell material not so much to earn but simply to eat. I am not investing anything right now, the only investment is my labor, nothing else. To sell brick at the price they want to buy them does not suit me."

It is instructive to probe more deeply into Tano's notions about brickmaking as a means of household sustenance rather than as a business. When I asked him why he was still making brick when most other plants were idle he replied:

> A lot of people ask me that question. I am working but really in brickmaking you don't earn very much. If I make a thousand brick and I sell them for 150, what I have left over is what I pay to another worker. I don't pay for firing, I don't pay for setting, I don't pay anything else. That is the source of my earnings. In order to keep on working, I invest only in clay and I tell my boys: 'Let's make three thousand brick today.' I don't pay them. I can work for a month without paying anybody.

Implicit in Tano's statement is that his plant viability strategy hinges on reducing labor costs in the form of wages to a minimum by relying on his own and his sons' labor in all stages of production from extraction and loading/unloading of clay, as well as cutting and/or loading of firewood, brick molding, and firing. The cost of firewood and clay would have increased by a factor of three if Tano did not supply labor. Tano's rent situation was more problematic, however, since the *ejidatario* who owns the plant passed away and his widow is not willing to accept rent payments in brick rather than cash.

In 1993 Tano, then age fifty three, was not particularly optimistic regarding the prospects for operating a brick plant according to this labor cost minimization strategy. This is how he expressed it to me:

> Look, I believe that I won't be able to do it any longer. I have worked a lot all my life and I am tired of working. I am going to stop doing it because there are no longer earnings with which to pay workers to keep them working. I still do it now because I am able to do it but I do a lot of work. I stay up at night in the brick plant and only come home to eat. That's why I don't think I'll be able to hold out much longer, maybe for another year or so. While I still can work, I will continue. A lot of people quit because there is no profit, and the people who stay have to be paid more.

In the context of expressing the view that his sons will probably not pursue their livelihood in the brick industry, Tano elaborated on the theme that brickmaking is physically punishing and takes its toll on those who do it:

> My sons are not inclined toward hard work. They work for a week or a month but not for years and years because it is hard work. This work has taken away young and old alike. Companions that I knew in Monterrey are now gone, and in Miguel Alemán, they no longer exist. It is a hard hassle. For someone who knows the work, it is easy to earn money. Knowing the work you can make money faster than in any other job. But as you get older you struggle more to make one thousand brick daily as you once did. You try to make more and

> more. It is a tremendous effort, pure back work, pure weight. If you don't get sick you quit simply because you can no longer do the work. You tire out and you can't do it any longer.

Aside from molding, stacking, setting, and loading/unloading brick, Tano as plant operator and boss was also responsible for locating and supplying clay that would provide the best possible quality of fired brick. During one of my visits to his *ladrillera*, he pointed to a nearby clay batch and said: "The clay that is piled up over there, we dug it. You have to look for it before you can dig it. We take samples of clay to find out which is the best. That is where we dig. When you see that good clay is no longer coming out you look for another place to dig. You bring a truckload of clay as a sample and see how it turns out."

Tano conducted an experiment that convinced him that the most important variable in making good brick was firing, especially in terms of intensity of the heat. He made three different batches of clay, using different soils and different amounts of water in the batch mixes. All of the batches yielded equally bad results under conditions of an overheated and under heated kiln, and more or less good results with a properly heated kiln. Tano concluded from his experiment that the matter of different clays or clay mixes was less important than firing temperatures in producing good brick. In his words:

> Too little heat or too much heat affects the brick. Depending upon the heat, brick will come out *sordo* or *chamuscado*, a bad material that has no sound [when tapped]. If the material comes out good, the heat is responsible. The clay needs only to be clean, without gravel, sand, or shells. All clay is equally good but it needs to be mixed to its point, that is, not too watery or soft and not too dry or hard.

Over the years, Tano has had a business relationship with Toño Reynoso that helped to keep his business viable. In the summer of 1993 I witnessed this relationship in action. Reynoso was short several thousand brick on an order from Houston and after loading at his own yard in Los Longoria he accompanied the truck driver to Arriaga's yard in Cavazos where he knew the needed brick were available. When I asked Tano about the relationship, he commented: "When they don't have enough to complete an order they come here to complete it."

On that same July day Reynoso picked up 8,400 brick from Arriaga to complete a trailer load of 13,316. The brick were unloaded from the kiln and carted by hand to the area where the trailer was parked and then were hand–loaded onto pallets placed on the trailer bed. The selling price was fifty–five dollars per thousand. If Tano would have had to deliver brick to Reynoso's ladrillera, he indicated that he would have charged an additional five or six dollars per thousand. The trucker–intermediary, Tiburcio Ramírez, told me at the time that he would resell these very same brick to a Houston dealer for one hundred twenty–five dollars per thousand. This illustrated in very concrete terms the dilemma of small producers like Tano—they are able to realize only a limited portion of the exchange value of the brick they produce. [See Appendix 1. Photos 23–33 show a variety of scenes and activities at the Arriaga plant, including the aftermath of a summer rain that ruined green brick drying on an open patio as well as the tractor–trailer loading process mentioned above].

On January 30, 2008 I found Tano at work in a small plant operating on land owned by Toño Reynoso in Los Longoria down the road from the new *Ladrillera Reynoso.* The brick that Tano and several other workers were molding on that day were contracted by the Reynosos and were to be fired in their kilns. Tano is now, in effect, a freelance brickmaker for hire who can no longer pay the exorbitant fifteen hundred dollars in annual rent demanded by the Cavazos landowner's widow from whose deceased husband he previously rented, at a much reduced rent, to operate his plant.

The Last Soto *Ladrillera*: La Paloma

Invariably, among *ladrilleros* the mention of the surname Soto evokes a reference to 'La Paloma' or 'doña Anita' who in the last decade of the twentieth century was known to be the last surviving representative of that renowned brickmaking family still active in the industry. As anyone who knows her will attest, Anita's near–celebrity status in the Reynosa brick industry also reflects her colorful and expressive personality, a lifetime of hard work in and sincere dedication to her grandfather's and father's craft, astuteness in her business dealings, generosity and perseverance as a *patrona*, and a straightforward, down–to–earth demeanor. In a 1996 interview she revealed that she saw many of these qualities in her father, Andrés: I remember my father clearly. He was an honest, honorable man who talked a lot. I am grateful that I turned out like him." She explained her own perseverance in the brick business as loyalty to the memory and craft of her father: "I am taking his part. That's why I remain hopeful about the brick plant."

This "*ilusión de la ladrillera*" is what brought Anita back into the brick business in the 1990s after being forced out in the early 1980s by the market decline. It boils down to a visceral loyalty to and love of the craft that she expressed in sentimental terms as follows: "The best life is at the brick plant, not in town. Go to a brick plant and there you find life is more tranquil, nicer. You look at things that are very authentic." There is no denying, however, that reality in the brick industry is often disheartening and frustrating. This was evidenced not only in the careers of her father and brothers but also in her own career.

After her father's death in 1964, Anita found herself obliged to get involved in operating a brick plant to help complete an unfilled order for twelve thousand brick her father's customer in Hidalgo. Texas. One of her brothers, who had been working with her father, made and fired brick for this client but since the clay had too much lime content many of them split or crumbled during firing. To make matters worse, it rained heavily during the firing. Consequently, only five thousand brick were salvageable from the kiln load of twelve thousand.

Confronted with this failure of her brother's firing, she decided to help out and rented a semi–finished kiln located on private property [*pequeña propiedad*] in Rancho Las Calabazas. Together with her niece, Irma, and her brother–in–law, Amado Garcia, who was to become her foreman [*mayordomo*], they dug a well, prepared clay, molded brick, and loaded the kiln with twelve thousand brick. Anita then went personally to her father's client in Hidalgo and told her that she was delivering the material in accordance with the contract her father had made.

She narrated that Alfonso Longoria of the nearby *hacienda* came by her *ladrillera* right after she rented it and told her he was impressed with her desire to work. She explained to him that she was doing so to uphold her deceased father's reputation and her family's name. Longoria was impressed enough to tell her to hire three workers and go to him the following Saturday to collect money for their wages. So, Longoria in effect financed her first three kiln firings. For many years she used the brand plate [*placa*] "*Ladrillera Soto*" and "Anita Garcia" but also produced brick as a sub–contractor for other *ladrilleras* with their own trademark frogs. For a few of the peak market years of the late 1970s she also operated a *ladrillera* in Rancho Grande.

It bears mentioning that Anita's younger sister, Andrea, also worked side by side with her in their family's plants. During the 1970s Andrea, in partnership with her husband Luís Domínguez whom she met in Anita's first *ladrillera* when he was a molder, had a plant in Camargo that molded and fired brick for Ramón Longoria.

During the mid–1960s, Anita either walked, or was given a cart [*carretón*] ride, to and from the family's house on Guadalupe Victoria street to her brick plant in Los Longoria. She elaborated: "We made material and we carted it. Amado Garcia was the foreman. We dug out clay in the pits with hoes and we wet it down in the afternoon." Her original financier, Alfonso Longoria fell ill and could not finance her beyond the three initial kiln firings. Fortunately, the market was by then on the upswing and the trucker–intermediary Luz López came to buy brick from her and she entered into her first of several sub–contracting relationships.

Until the mid 1980s, when she withdrew from the brick business in the aftermath of the "big collapse" of the market [*el gran desplome*], Anita mostly operated her plant to produce under her own brand for trucker–intermediaries or for the large semi–mechanized ladrilleras under their brand name. One of her first arrangements of this type was initiated with Luz López in 1974 when he drove up to her *ladrillera* and told her: "I'll advance you money for the worker's wages." For four years after that first meeting, according to her, "All of my output was for don Luz López—brick of eight, nine, and ten inches. He bought to sell in Texas."

Next, Anita entered into a business relationships with Elizardo González, who had a trucking business but also had kilns not far from Soto's *Ladrillera La Paloma*, and with several others including Mario Morelos of *Ladrillera Morelos* and Joel Rodríguez of *Ladrillera Reynosa*. She explained these relationships as follows: "I produced unfired brick with clay and molds they provided. I put the brand plate [placa] of don Elizardo. I made brick for Morelos and also for *Ladrillera Reynosa*." But as these large *ladrilleras* expanded their output through opening satellite fields directly under their control, they offered to pay less for unfired brick from Soto's plant, thus, forcing her to reduce her business with them. In the late 1970s, however, the Varela's sub–contracted Soto's plant to make unfired brick for their Cavazos plant [*Ladrillera Industrializada Rio Bravo*] with clay supplied by them. During this period she was producing fifteen thousand to twenty thousand brick weekly with an average of ten molders [*hechureros*]. She explained: "They advanced me the wages for my workers, the brickmakers. I put my truck, what I wanted to do was to earn from my truck. I transported material to their plants so that my truck would earn." In short, the only way Soto's plant could stay in business during this period was as a satellite of larger export plants, and her limited earnings were mostly derived from her role as a hauler rather than as a producer.

Elizardo González confirmed having done business with Anita Soto who he referred to as "the famous La Paloma," during the late 1970s and into the 1980s. He also did business with Andrea, Anita's sister. He claimed that both sisters still owed him money, but had no regrets because business was good in those years. This suggests that, given the low profit margins under which the Sotos were forced to operate, they had to deliver short as much as possible in order to survive. Large volume intermediaries like González, were making money from high volume export sales and simply had to assure their capacity to supply brick through the campero arrangement. They could easily absorb shorted deliveries of unfired brick so long as the deliveries kept coming. González seemed to acknowledge this when he admitted: "When the time came that there was no market, I stopped taking more brick from the Sotos." That was also the cause of their alleged shorting of his deliveries. But he emphasized, in closing, his recollection that "they brought together a lot of people . . . and over the period that we worked together, they had very steady output."

Anita's following reflection on the shutdown of her plants in the 1980s highlights the importance of the business relationship between small and large ladrilleras in Reynosa but reinforces the interpretation made above: "We had nowhere to sustain ourselves. We kept going because *Ladrillera Reynosa*, *Ladrillera Morelos*, those people were sustaining all of us. If they stopped doing business, well we also stopped. They were the ones that gave us orders for material. They were the ones that supplied us with clay in order to work. Without them we were unable to do anything."

Regarding her personally painful decision to shut down her the *ladrillera* that was producing mostly for the domestic market, Anita assigned blame to the trend toward cement block in the Reynosa construction industry. As she explained:

> I stopped when I saw that the others stopped. I stopped and it was sad. There was no one. Everything was dead. Dead. All of us came to the end. All of us stopped. No one wanted material. The entry of cement block into the market ruined us. No one wanted to buy adobón. The architects and the engineers began to buy cement block from Monterrey and they completely ruined us. The brick industry ended for all of us. It became a memory.

Fortunately, contrary to Anita's bleak assessment, there was still a little bit of life left in the industry.

The several years in the 1980s after her plant shut down encompassed the only time during Anita Soto's adult life that she was uninvolved in brickmaking. Although she and her husband also had a waste cardboard recycling business they had started in 1955, she had herself also been in the brick business until the 1980s plant shutdown. Afterwards, she spent some time in Dallas as a tamale–maker for intermediaries who sold them to the African–American population. She made 150 dozen tamales weekly and sold them for $7.00 per dozen. Soto's earnings from tamales, together with those from the cardboard recycling business, helped her educate her three sons, make a down payment on a used truck, and to eventually reopen her Las Calabazas *ladrillera.*

Anita's return to brickmaking in 1991 was apparently greeted by skepticism from the *ejidatarios* from whom she rented the site. In her words: "They thought I was crazy and asked me who was I going to sell to." She elaborated on her thinking at the time: "When I saw all of the kilns tumbled down, it made me sad. I knew that I was willing to work. I knew that I could get the plant operating. I was not too hopeful. I knew that some day, God willing, that the plants would be operating again. I had to do it." How did Soto put this plan into action? She explained: "First, I dug a well. Then, when we had water, we brought the clay. Then I bought a used truck, so I was paying for it and repairing the kiln. I had a foreman who brought two molders. In order to pay my workers I collected and hauled three truckloads of cardboard to sell on the other side. That's how I started the plant up again."

It was this reopened plant that Anita chose to identify with the brand name La Paloma, a nickname she had been given over the years through her husband who had been called *El Palomo* since boyhood owing to his custom of dressing in white clothing. Few knew the origin of her nickname but everyone in the brick business knew the identity of *La Paloma.*

Ironically, when she re–entered the brick business in Calabazas in the early 1990s, the market situation was not much better than it had been when she shut down in the 1980s. A small plant like hers had to depend primarily on sales in the local construction market supplemented with occasional export sales to trucker–intermediaries. Only two of the larger gas–firing plants were still in business in the 1990s and neither was operating at capacity; consequently, they had no need to sub–contract production to small plants. Thus, after her reopening all of her sales were either to building materials suppliers in Reynosa or to an intermediary–trucker named Padilla from Matamoros who resold in the greater Matamoros market but also in Texas.

Accordingly, when I interviewed her in 1993 Anita painted a dismal picture of the overall market situation:

> There are no sales. The business is dead. The client in Matamoros I supply takes a load every two to three months. He buys on credit. This week he came and took ten thousand [ordinary brick] and three thousand *barras* [long brick] to complete a load. So, it is very difficult to make it here. I sell to the lumber yards and hardware stores here to be able to keep the business going. The U.S. market is dead.

At that time there appeared to be some renewed interest in a brick she described as "a thicker nine, the old nine inch that measures nine inches in length by two–and–one half inches in thickness." It was being made at the request of the Matamoros client described above, who also bought ten thousand "*ochitos y medio*" (8–1/2 x 2 inches). The *adobón* [11–1/2 long x 6 inches wide x 2–1/2 inches thick], was being made as well to sell locally. *Adobón* remained the preferred brick for the Reynosa construction market in the early 1990s but its use has subsequently diminished.

During the mid–1990s *Ladrillera La Paloma* had three or four employees, several of whom lived in housing at the *ladrillera* with their families, and a foreman who also was in charge of firing the kiln which had a capacity of forty thousand *adobones*. The employees molded brick, stacked it after drying, set unfired brick in the kiln, unloaded and restacked fried brick from the kiln, and assisted in loading and unloading trucks. One of these workers, Ernesto, was originally from an *ejido* community in Querétaro but had been in the border region for twelve years, first in Matamoros where he learned brickmaking and afterwards in Reynosa. He had been working in *Ladrillera La Paloma* for only two months when I interviewed him in 1993. In comparing agriculture with brickmaking, he felt that "it is the same regarding the punishment one suffers but planting is nicer, agriculture is nicer." Another worker, Constantino, from a small community near Linares, had been working for La Paloma for a year and a half, and had been in Reynosa for some fifteen years. He worked first in agricultural and ranch work, and then as a mason's helper [*peón de albañil*] before working in brick plants. He preferred brickmaking because he had to work fewer hours daily than in other jobs.

José Luís Martínez was born and raised in *Ejido Los Longoria* or as he expressed it, "I made myself into a man here" [*Me he hecho hombre aquí*]. He had just started as the foreman [*mayordomo*] and stoker [*quemador*] at *Ladrillera La Paloma* in 1993. His father was born in 1901 in a ranch near Monterrey but came to Reynosa in the 1930s, settled essentially as a squatter [*paracaidista*] in Los Longoria with the intention of using it as a base for migrating to the U.S. He did manage to work as a *bracero* in the 1940s but also learned brickmaking in Los Longoria. José Luís was born in 1955 and at the age of twelve started working with his father in a local *ladrillera* owned by Rafael López de la Cruz. In his words:

> My teacher was my father, it was him who taught me. We made six thousand pieces daily between us. We did everything right there. There was everything, including a kiln. He had large clay pits with very good clay to make material. Mr. Lopez de la Cruz was one of the pioneers who began on a small scale with a little kiln, and then another, until he came to have six kilns. There were about fifteen of us who worked with him. I worked for him for about twelve years.

During those twelve years in the López *ladrillera* [*Materiales López*] José Luís became proficient in brickmaking and in firing: "We monthly filled a kiln of 90,000 pieces and yearly we completed some six or seven kiln loads. I molded brick, lighted the firebox, and fired—it took seven days to get good brick." He noted that because of cold or wet weather, several weeks of work annually were lost—a factor that continues to plague the industry to this day, even when brickmaking patios are roofed and protected from rain.

When López de la Cruz died in the early 1980s, José Luís went to work for about three months with Anita Soto but when she closed down her plant, he went to work for some local ranchers doing maintenance work on the irrigation system, cutting alfalfa, and other odd jobs. When Soto re–opened her plant in the early 1990s José Luís came back to work for her as *mayordomo*. In the summer of 1993 he was making an average of one thousand brick daily, supervised five other workers, kept a count of weekly output, and was also responsible for setting and firing the kiln. [See Appendix 1. Photos 41–46 show various scenes and activities at *Ladrillera La Paloma*. José Luís Martínez appears in photo 47].

As José Luís and I walked and drove through the extensive brickmaking zone of Los Longoria–Calabazas in 1993, he expressed bitterness and frustration about the conversion of brick plants into trash dumps, a fate that affected most of the area's small plants including that belonging to the López family. In walking by the place where the López plant had once thrived but in 1993 served as a dump for huge mounds of trash, José Luís commented, "They ruined the earth by tossing all that garbage there." Then he gave his explanation of the transition of that brick plant: Mr. Sanchez's sons worked there and made brick. But they rented the plant to a person who put a trash dump there. That person began to allow cartfuls of trash to be dumped there. Within a year Sanchez's sons, seeing that the renter had developed a good business, decided to no longer rent to him and they took over the trash dumping business themselves. The result, according to Jose Luis, was that they ended–up "killing the earth" [*Acabaron de matar la tierra*]. He admitted however that there were advantages to the transition since the trash haulers simply come there to dump their carts. They get permission from the owners to do so and pay no fees. The owners then have the right to sort the trash—separating out aluminum, copper, iron, cardboard, and so on. They do not have to pay and supervise workers as they had to do in their brick plant. After admitting all of the above, and after walking past the site, José Luís muttered "they have a tremendous pigsty there that is hard to tolerate."[3]

The Coahuila Nexus

In response to my question as to where she got her workers during the boom years, Soto explained that when the Valeras had labor difficulties in their *Ladrillera Industrializada Rio Bravo* in Cavazos several of the disgruntled workers came looking for work at her *ladrillera.* They were from the area of Torreón, Coahuila and told her that if she wanted to hire more workers, she should go there—particularly to Mayrán near San Pedro Las Colonias. She followed the advice of these workers and went to Coahuila. The conditions she found there made a lasting impression:

> Mister, it made you sad. Those poor people! If you go there now in your car and make a public announcement that you need so many workers, a large number of people will show up—teenage boys with the hope of getting work in order to eat. There are brick plants around there but they don't provide employment. There is no activity, no life. And the agricultural fields there, well, there was no source of payment. It was very sad. I'm telling you, sir, it was pitiful. All of those people.

On the basis of information from Anita Soto and others regarding the participation of workers from the greater Torreón area of Coahuila in the Reynosa brick industry during the 1960s and 1970s, I traveled there in the summer of 1995 for the specific purpose of visiting the city of San Pedro de las Colonias [population around eighty five thousand in 1993] and the small settlement of Mayrán to find out more about this topic. On the road from Torreón, just to the west of the urban center of San Pedro, I visited an area of *ladrilleras* with several *ladrilleros* at work. Clay pits and the molding/drying patios were close to the kilns that had capacities in the range of twenty seven to thirty five thousand pieces. At the time three sizes of brick were being produced—small, medium, and large that resembled the Reynosa *adobón*. The selling prices per thousand of these brick was from 100,000 pesos [$32.00] to 300,000 pesos [$97.00] according to size and the molders were paid from 20,000 pesos [$6.50] to 60,000 pesos [$19.35] per thousand also according to size.[4]

Several workers I talked to in San Pedro Las Colonias and Mayrán had participated in migratory labor in the 1970s and had experience working in *ladrilleras* in Ciudad Juárez and Casas Viejas in the state of Chihuahua, as well as in Ciudad Miguel Alemán and Reynosa in Tamaulipas. One of them expressed the reasons for this participation in migratory labor as follows: "When work here was hard, when there were no sales and worked stopped, we left for elsewhere" [*Cuando estaba aquí muy duro el trabajo, cuando ya no había ventas y se paraba el trabajo, salía uno para allá*]. Another stated that many people left San Pedro for Reynosa in the 1960s because "There was nothing else to do. Things were very bad, the weather was poor. The reservoir dried up, there was no water." [*No había que hacer. Estában muy malas las cosas, muy pobre la estación. Se agotó la presa, no había aqua*] According to these workers, the local brick industry received a boost in San Pedro in the 1960s by returning migrants who had worked in Reynosa. The original plants are near the cemetery, not far from the town center, but are relatively inactive since land was set aside by municipal authorities for brickmaking outside of town—precisely in the area where ladrilleros were working at the time of my visit. [See Appendix 1. Photos 48 and 49 show the newer plant area in the desert west of the city, and photo 50 shows the original *ladrillera* in the urban area near the cemetery].

The theme of drought and scarcity of water for crop irrigation also emerged in my visit with officials in the small *ejido* community of Mayrán, referred to by one of them as a "*congregación*." When I asked them about migration from the *ejido* to Reynosa, a response that everyone present seemed to agree with was "Many people, have to leave to find work elsewhere due to small harvests, many people are unable to plant due to a lack of water for irrigation" [*Mucha gente, pues, tenemos que salir para trabajar afuera por poquita cosecha, hay muchas gentes que no pueden sembrar por falta de agua*]. Both San Pedro and Mayrán are within the Irrigation District of the Laguna Region [*Distrito de Riego Región Lagunera*] in which the largest reservoir is the *Laguna Mayrán* but only *ejidatarios* have access to water, so brickmaking is basically the occupation of offspring of ejidatarios who do not have water rights [*ejidatarios sin derecho a agua*] or to non–*ejidatarios* or so–called "free" [*libres*] members of the community.

In 1995 Mayrán had only fourteen heads of household who paid a fee to the ejido for rights to operate a *ladrillera*. Each of these *ladrilleros* employed several family and wage laborers. Many of these plants operated year–round, whereas others are more seasonal in operation. Kilns tended to be smaller than in San Pedro but use the same fuel, a combination of discarded tires, recycled pallets, firewood, and dried animal manure [*estiércol*]. Firewood, however, was becoming increasingly scarce in 1995 as was apparent from an ejido official's statement that the *ladrilleros* "struggle to obtain firewood" [*Batallan con la leña*].The *ejido* of Mayrán had 700 heads of household in 1995 and the consensus of *ejido* officials was that the majority have worked in brickmaking. In addition to Reynosa where two colonias, Rancho Grande and Cavazos, were cited, they acknowledged that many from their community had worked in *ladrilleras* in Camargo or in *Ladrillera Falcón* in barrio Los Angeles, Ciudad Miguel Alemán. Also, tractor–trailers were said to have arrived in Mayrán and San Pedro in the past to load brick, indicating that there was export production, possibly linked to Camargo and Miguel Alemán and, also, to Piedras Negras and Nuevo Laredo.

Conclusions

Obviously the closing of brick plants starting in the early 1980s was accompanied by corresponding reductions in the labor force and, in fact, by a permanent reduction in the industry's labor pool. "There are fewer employers in the brick industry nowadays," lamented Anita Soto in 1993, "and fewer workers." Before the 1980s in periods of market slump and layoffs, *ladrillera* workers would usually find work in the construction industry at lower wages than were paid in the brick industry. When the brick business picked up they would return to work in the *ladrilleras*.

An increase in demand for Mexican handmade brick in the early months of 1999 was substantial enough to re–initiate, albeit on a much reduced scale, the satellite field strategy in one of the two gas–firing *ladrilleras*, as well as intensify competitiveness among the *ladrillera* operators for available brickmakers, especially the small–scale plants which are more prone to labor turnover. These plants, with only unsure access to casual laborers and a minimal pool of dependent household labor, are the most economically squeezed units in all market conditions. Recently onesuch plant, owing to a scarcity of labor and cash, agreed to sell its entire output of crude brick to a larger gas–firing plant for a price of $45.00 per thousand. It was more favorable for the worker/owner, Tano Arriaga, to do this rather than to fire his own brick and then wait for buyers—who probably will not pay more than $60.00 to $65.00 per thousand for fired brick. In his case, however, this proved to be a short–lasting solution to viability since his profit was so reduced that he was no longer able to meet the cost of rent for his plant. Subsequently, his only alternative has been to hire himself out as a brick molder to plant operators like the Reynosos.

The wood–firing, small–scale *ladrilleras* covered in this chapter were either on rented *ejido* land, on land held under titled possession that, for all practical purposes, was recognized as private property, or on land held as private property. They were operated by individual *ladrilleros*, with or without the participation of family workers, but often with one or more hired hands, all of whom either lived at or in close proximity to the plant. During the 1980s and 1990s most of these plants produced *adobones* primarily for the domestic market, selling small lots of one thousand or fewer brick to local buyers for own use, although sometimes selling larger quantities to building supply companies or to masons through informal contractual arrangements. These plants, as micro enterprises, on occasion paid minimal taxes but did not pay for social security benefits for their hired hands. All of the above features place these plants in a similar category to those covered in my Oaxaca study (Cook 1984), those visited by me in Ciudad Juárez, or to those studied by Wilson (2005) in Mexicali. Given an apparent trend in the Reynosa masonry construction market away from clay products and increasingly toward the use of cement blocks, the continued viability of this sector of Reynosa's handmade brick industry may reside on their capacity to participate in production for the Texas market, either through relationships with bilingual trucker–intermediaries or with the few remaining export plants as suppliers of unfired brick.

Notes

1. The other gas–firing plant was *Ladrillos Reynosa* operated by Minerva Morales, widow of Gustavo Garcia. Due to their histories as rural areas with low population densities, all of the colonias are also *ejidos* or agrarian reform communities where the land is either in the hands of (1) inhabitants who were granted use rights by Mexico's agrarian reform bureaucracy as *ejidatarios*; (2) another group of migrant settlers who, as renters or squatters, have been granted rights of possession [*posesionarios*]; or (3) private property owners who had or have ranches, lots or other landed property [*pequeños proprietarios*].

2. These were operated by the following persons: Anita Soto, Guadalupe Garza Balboa, Francisco "Pancho' Macías, Martín Guajardo, Antonio 'Toño' Reynoso, Jose Luís Vigil, Refugio Dávila, and José María Cantú.

3. In her study of Mexicali Wilson (2005:117) found that brickmaking and garbage picking were among the six most common occupations of male heads of households in the settlement she called *Colonia Popular*. She provides a detailed analysis of the garbage–picking families and their activities (2005: Ch. 9).

4. These product prices and wage rates were provided by informants in old pesos not New Pesos. The official change in the Mexican monetary system occurred in 1993 (one New Peso was equivalent to 1,000 old pesos). At the time of the change, the peso–to–dollar exchange rate was approximately 3,120 to 1. Apparently, Coahuilean *ladrilleros* in 1995 still preferred to use the old system to calculate values. It was unwieldy but still familiar and, apparently, old pesos were still in circulation at the time. I calculated the dollar equivalents on the basis of the 3,120 to 1 rate. I did not witness or obtain documentation for any transactions in Coahuila to confirm these volunteered price data so they must be interpreted cum grano salis.

San Pedro de las Colonias and Mayrán are located in the Upper Basin of the Nazas River, also known as the Region of the Lagoon Basin [Región de la Comarca Lagunera]. The Rio Nazas originates in the mountains of Durango, and then flows eastward to the Laguna district where during the rainy season it would reach and disappear into the

Mayrán Lagoon. The natural flow of the river has been considerably reduced by the Zarco and Cárdenas dams in Durango, and only in unusually high rainfall seasons did its waters flood the Coahuilan desert basin and reach the Laguna de Mayrán. Apparently, this has not happened for quite some time. I did not travel to the Lagoon that begins at the village of San Nicolás about 6 kilometers east of Mayrán by unpaved road, so I have no first–hand knowledge of the condition of the Lagoon in that year.

Chapter Eight
Survival of the Fittest: Gas–Firing *Ladrilleras* and Cross–Border Truckers

The introduction of gas–firing restricted the location of plants that chose to use it to as close to the gas line as possible in order to reduce installation costs of the delivery infrastructure. Especially in the case of the earliest gas–firing plants, this resulted in locations that were already urban or in the process of becoming so, and that were farther from the river and clay sources than were the locations of the wood–burning plants. The adoption of gas–firing also went hand–in–hand with semi–mechanization and an increase in plant size as economies of scale thinking became more relevant. Clay was either trucked to *ladrilleras* to be pugged and molded into unfired brick, or unfired brick were made in satellite fields in the riverine zone and then trucked to the main plants for firing. As urbanization continued, the *ladrilleras* were moved to locations on the gas line on the periphery of the city, but still with quick access to main roads.

The last two of these gas–firing plants still in operation in 2000, namely, *Ladrillera Reynosa* and a branch of *Ladrillerías Camargo*, *Ladrillos Reynosa*, are located side–by–side in *Colonia Rancho Grande* just off the riverine highway to Nuevo Laredo which runs in an east–west trajectory immediately to the south of the Anzaldúas Canal. Map 2 shows the general location of these plants as a black dot just above the white lettering for Rancho Grande. This is just south of where both the Anzaldúas canal and the riverine highway turn northward.

'Made in Mexico to Last': *Reynosa Brick Company*

The most active of these plants with more than forty years of participation in the Texas market is Joél Rodríguez's Reynosa Brick Company [*Ladrillera Reynosa S.A. de C.V.*] whose logo includes the phrase "Made in Mexico to Last." In 1992, according to René Ramírez, plant supervisor, the plant had sixteen molders making sixteen thousand to twenty thousand brick daily from Tuesday to Saturday. Each week's production required ten to fifteen days to dry [*la secada*] prior to being loaded into the kiln for firing. Ramírez reminisced about his experience in the plant as follows: "I have been working here for thirty two years, the same as the age of the plant. About twenty years ago a lot of brick were sold, one hundred eighty to two hundred thousand—about eighteen trailer loads weekly. About six to eight years ago, sales declined. Now between four and seven trailer loads weekly are sold; each trailer carries around eleven thousand brick."

During the 1970s and early 1980s *Ladrillera Reynosa* was exporting an average of eighteen trailer loads of brick weekly to Texas, primarily to Houston, whereas during the 1990s it exported an average of three or four trailer loads with a staff of twenty–five workers. It has the distinction of being the only large–scale, semi–mechanized, gas–firing plant that has remained continuously in operation since the 1950s. Under Rodriguez's ownership and management since 1960, clay preparation, kiln–loading/unloading, truck loading have been mechanized, and other improvements made to increase labor productivity and overall efficiency of the plant's operations. It is also the only Mexican *ladrillera* on the lower Rio Grande border that, over the decades, has established brand–name recognition as 'Reynosa Brick' in the Texas market through its dissemination of its attractive English–language brochure, distributed mainly through Texas importers/dealers, that has informative text, colored photos, and tabulated laboratory test results for its buff, goldtone, *rosillo* and light pink "natural" clay brick. The company's line also includes thirteen additional "colored" brick ranging from white [*blanco*] to red [*rojo oxido* and *rojo regular*] to grey [*iglesias gris*]. The average size of these brick varies from 7–3/4 x 3–1/4 inches x 2–3/8 to 9–7/8 inches x 3–1/4 x 3 inches in length, width, and height. Since the 1970s, this plant has combined traditional hand molding with mechanized technology, including machinery that moves and mixes raw materials for consistent clay texture and composition, and ten gas–fired kilns with capacities ranging from 95,000 to 160,000 brick each that reach firing temperatures in excess of 1700 degrees F.

The Mexican handmade brick export industry, according to Joel Rodríguez, has always been a quasi–dependency of the mechanized Texas brick industry serving the Texas market as a reserve source of supply when, for whatever reason, machine–made brick is in short supply. In rejecting the idea that these periodic surges in demand reflect a change in buyer preference, he reiterated the reserve supply source argument with a global system twist: "There is a market need and if U.S. industry can't satisfy it then industry in some other area of the world will. In that case we Mexican suppliers are the first in line because we are closer . . . This is a pattern that has been repetitive."

In the late 1990s, as the largest of Reynosa's two remaining gas–firing, semi–mechanized brick plants, *Ladrillera Reynosa* had a plant layout and work schedule characteristic of all the large plants of the past century. When kilns are in the process of firing, the plant operates day and night until the firing is completed. Most other operations like clay preparation and molding are conducted during daylight hours, six days a week year round, although schedules are adjusted to weather and market conditions. The plant has three large kilns with capacities of two hundred thousand pieces, and nine smaller kilns with capacities between ninety thousand to one hundred thousand pieces, all gas firing. Much of this kiln capacity, however, has either not been used or has been used only occasionally since the 1980s; several kilns are no longer serviceable and are in need of repair. The kilns used most often in the 1990s are located in the eastern half of the plant site, on either side of the truck loading area, and just to the north of three molding/drying sheds. [See Appendix 1. Photos 51–62 show the facilities and activities at this plant].

In the 1970s this plant was selling an average of 180,000 to 200,000 brick weekly amounting to eighteen to twenty trailer loads. In the 1990s the average declined to an average of forty thousand to seventy thousand pieces or four to seven trailer loads weekly. For example, the reported output, measured in trailer loads of approximately eleven thousand to twelve thousand brick each, for the first half of 1993 was: nineteen in January; twenty in February; twenty-seven in March; twenty-seven in April; twenty-six in May; thirty in June; and fifteen trailer loads had left the plant as of July 15. On the latter date, the plant had an estimated inventory of eight hundred thousand to one million brick in the process of drying in the sheds, six hundred thousand unfired brick set and ready to fire in the kilns, and another 450,000 fired brick palletized and ready for shipment. My visit to the plant at that time showed that the sheds were essentially filled to capacity with molded brick at various stages in the process of drying.

Even under these conditions of reduced sales and output, chances were that on any given day of the year one or more kilns was firing and at least one was being drawn or unloaded to supply existing market demand and inventory maintenance. Once a kiln is set or loaded, the firing process takes a minimum of one week and the cool–down at least another week. Setting or loading the kiln can vary from a few days to two weeks, as does the drawing or unloading process. This means that a complete cycle of setting, firing, cool–down, and drawing of kilns will take a minimum of thirty days and possibly as long as forty–five days. This is why the night–time visitor to *Ladrillera Reynosa* will encounter not only a watchman but also one or two workers who are tending a kiln in the process of firing (see Cook 1998:190–191 for more details regarding kiln cycles and sequencing).

On a typical work day during the week there is usually activity in all sectors of the plant site: clay is being milled, mixed, and delivered in batches of mud to the molding floors; brick are being molded and dried; unfired, dried brick are being palletized and transported to kilns for setting; at least one kiln is being drawn or unloaded; fired brick are being slurried or colored; and flatbed tractor trailers are being loaded with fired brick headed for the Texas market. In short, the casual visitor and the ethnographer alike are presented with an array of different activities to capture their attention.

Entering the plant site from the northeast corner where the office is located, immediately to the west is the first block of three kilns followed by a second block of three kilns on the west side of an open truck loading area. Directly to the south, side by side, are two equally imposing elongated, open, roofed sheds known as *casilleres* or *secaderos*. Each shed houses twenty–five separate patios where brick are molded, canted, and stacked in towers until completely dried, and then palletized and stored until transported into the kilns by forklifts for setting and firing. Three additional sheds of equal size are located in the same sector to the west.

A principal reason why these buildings are so large is to accommodate enough roofed space for molding brick in a moist state that requires them to be laid out [*tendido*] to dry, a process that occurs every day under normal circumstances. The drying process involves canting and then stacking brick in towers until they dry out. Rain and humidity, as well as the content of riverine clay in the Reynosa area, inhibit fast drying. Consequently, the molding and drying of brick requires a sizeable covered area and takes anywhere from ten days to two weeks. Separate from but near to the molding/drying sheds is another open, roofed building housing the clay mill and mixer, as well as a large storage area for raw and milled clay. Mixed clay is loaded directly into dump trucks that deposit it at the appropriate molding areas.

In July, 1993 I spent time with José Angel Agüero at his molding station and watched him mold as we talked. At his molding station, a few feet behind him was the clay mound [*soquete*] that had just been deposited by the plant's dump truck. He referred to the overall task of molding brick as *echar ladrillo* [literally *to throw* or *toss* brick] and broke this process down into the following steps: sand the mold [*arenar el molde*] to prevent the wet clay from adhering to the mold at the time of release onto the drying floor, and then fill the mold with moist clay [*llenar el molde*]. The latter step itself involves two sequenced activities: a) making a cone [*hacer cono*] or making a small cone–shaped batch of clay from the larger mound, and b) cutting a little ball from the cone to throw or toss it into the mold [*cortando la bolita para echarle*]. The task of cutting [*cortar*] clay is performed strictly by the hands of the molder, without the aid of any type of tool. This is followed by the final task of scraping [*raserar*] excess clay off the top of the mold with a wooden scraper [*rasera*].

This complete sequence of tasks is performed at a fast pace with the molder on his knees, with the mold on the ground directly in front of him, and a pile of sand to his side. When I commented to José that he performed this tasks very rapidly, he responded that "others toss two, I toss only one" ["*Otros echan de a dos, yo echo de a uno*"]. He was referring to the fact that he tosses one clay ball per mold compartment or for each brick in the mold, whereas others cut a two brick–sized ball or one that fills two compartments in the six compartment mold. I confirmed this in observations of other molders at the plant. As of the end of August, 1992 the plant had six full–time brickmolders, and ten others who sometimes also molded. The total output was between sixteen thousand and twenty thousand brick daily or eighty thousand and one hundred thousand brick per week.[1]

One notable result of mechanized clay preparation at *Ladrillera Reynosa* has been to substantially increase daily worker productivity simply because it saved them the drudgery of having to cart raw clay from the main supply point, haul water, and mix the clay with a hoe. Before mechanization the typical molder reached his physical limit of output at one thousand brick daily, whereas after mechanization that limit has risen to between fifteen hundred and two thousand brick daily or ten thousand brick for a five–day week. During the 1990s, the company has adopted a target [*meta*] output system with a sliding pay scale that pays the highest piecerate to workers who meet the output target, and a lesser piecerate to those who produce less than the target. It discourages output above the target and does not increase the piecerate for extra–output.

Dependent upon managerial decisions weighing current inventories against future orders, the output targets for different sized brick in the 1990s fluctuated from weeks with no output to some weeks with five thousand pieces per molder targets, and other weeks with a target of ten thousand pieces. The output targets were set as many weeks in advance as feasible. For example, when I visited the plant in November 1992, the last target set for eight–inch brick was nine thousand per molder per week which was paid at a rate of 27.280 pesos [$4.70] per thousand for the molders who met the target, or a weekly salary of two hundred forty–five pesos [forty–two dollars]; the target output for nine–inch brick was eight thousand at a piecerate of 28.800 [$4.95] per thousand; and the target for the large so–called bar [*barra*] was seven thousand at 31.900 [$5.48] per thousand. This was a relatively high production quota considering that René Ramirez estimated that the plant had a inventory of some three million fired and unfired brick. Then, again, it was not uncommon for the plant to receive single orders for five hundred thousand or more brick.

As can be inferred from these numbers, the piecerate per thousand pieces increased as brick size increased. The logic here, according to the plant's secretary/bookkeeper, Olga Magallán, was because molders prefer to produce either eight–inch or nine–inch brick and only a few of them like to make larger brick or specialty brick because the molding process is either more physically exhausting or entails more tasks. "Some of them are better at molding eight–inch brick, others the bars, and still others the nine–inch brick" she elaborated. As this rate differential implies, not all molders are equally inclined toward or capable of molding the entire range of brick. Ramírez summed up the situation: "Not all of them work the same. Some of them are very good; others are just average. Some are more skilled than others at brick molding."

In times of production slowdowns or stoppages, the plant's policy varies according to the underlying cause. When bad weather like rain or cold is responsible, the company suspends payment but does continue to pay benefits and offers limited pay advances. If, on the other hand, the stoppage is temporary due to equipment breakdown the company pays the less than target output at the weekly target piecerate.

Most of the molders I talked to were not enthusiastic about the target output system of payment, mainly because the target was often less than they were prepared to produce and earn. Although appreciative of the plant policy of providing wage advances or loans during periods of no output, as well as the continuation of social security benefits, all of them viewed life under those circumstances as being hard and found it necessary to obtain supplemental income. The plant management is well aware that workers have great difficulty in making ends meet during production slumps and attempts to compensate by allowing specialized workers to earn pay for doing other jobs in the plant. For example, molders can earn extra pay for loading trucks, setting and drawing brick from the kilns, coloring brick, and so on.

The case of Hermilo Ramos Castañeda is illustrative. Born in 1937 in San Juan Matías, Coahuila, he came to Reynosa in 1955 and found work for three years in a La Copa plant that was exporting brick to Louisiana by train from Hidalgo, Texas. He also worked for a year in agriculture, spent time as a setter in a Ciudad Victoria plant, was a brickmaker with *camperos* in Colonia Corrales, and worked three or four months in the Tex–Mex plant with Omar Zapata and Antonio Méndez. After 1960, he worked mostly for Gustavo García and, later, for Joel Rodríguez at *Ladrillera Reynosa*. He remembered the 1960s as involving more work for less pay than the 1990s but also admitted that there used to be more molders then because there was more demand for material. He has a house and residence near the plant site and said that he couldn't really expect more in life given only three years of schooling. Since the plant mechanized, he has produced as many as 10,600 brick weekly but admitted that now "we don't make much effort to make so many. We 'drag by our balls'" [*guevoniamos*] around here. He elaborated:

> Well, the task [*tarea*] is five thousand. That is how many we have to make, or they pay us at a cheaper rate. That's the way it is. When we are really working, we are tasked with eight thousand or ten thousand. Not so long ago we were stopped for quite a while. We worked a few days and then we were stopped. We were at rest for five weeks. We did not make anything. We did not earn anything. That's when they lend us eighty pesos a week. What can we do about it? Not anything to be proud of. That's the way it is.

Hermilo proceeded to reminisce about the 1960s when he earned four hundred pesos [thirty–two dollars] weekly as a kiln setter: "That was a lot of money [*un dineral*] then and went a long ways [*rendía*]. With one hundred pesos [eight dollars] one could buy a week's worth of food and you had the rest to spend as you wished. Today, by contrast, what you earn here is not enough to buy food." With a production quota of five thousand brick that was just introduced the week that I interviewed him, Hermilo said that it does not give him enough income to pay household expenses. Fortunately, he still has a son and daughter who contribute to the household income, and he does day jobs in construction to earn extra income when there is not enough work at the brick plant.

Manuel Varela, a native of Linares, Nuevo León, came to *Ejido Rancho Grande* with his father in 1947. His father had a small dairy farm and Manuel milked cows, and then worked in agriculture. He plowed fields with an ox team and mules. In 1963 he started working for Joel Rodriguez and Rene Ramírez at *Ladrillera Reynosa* as a kiln loader/unloader and clay hauler [*carretillero*] before the plant motorized. He also plastered over the kiln openings to seal them prior to firing. When the plant purchased forklifts, he became a driver [*motocarguista*]. He worked by contract, fifty pesos [four dollars] per kiln, and often made as much as two hundred pesos [sixteen dollars] per week that was more than enough to meet his weekly household expenses of fifty to eighty pesos [$4.00 to $6.40]. He has raised four sons and has a house and residence lot in the center of Rancho Grande thanks to *ejido* authorities. Manuel had little schooling and, for that reason, does not permit his sons to work at the brick plant, requiring them instead to complete their schooling. He wants his sons to have a better life than he has had. "I started to work when I was very little, taking care of cows, all the time, and I quit school. If I would have gotten more schooling, I would have gone to the other side [United States] or to another place."

Given his agricultural background, Manuel was able to gain access to some land from an *ejidatario* through an informal rent arrangement according to which he raises a variety of irrigated crops including corn, squash, watermelons, chiles, and tomatoes, as well as animals like cows, pigs and chickens strictly for household consumption and, in return, he helped out with ranch work. One of his sons also helped with the ranch work; another son worked in a *maquiladora* earning five dollars hourly and helped out with household expenses. Two of his sons were teenagers and full–time students.

Manuel has a five and one–half day work schedule at the plant from 7 a.m. to 4 p.m. but this varies according to market and weather conditions. There was much less work at the plant in 1993 than there had been in the 1960s and 1970s. The week prior to my interview with him he had loaded five trailers. This was more or less a typical week, although he said that he had loaded as many as ten trailers a week in recent months.

The work histories of piecerate brickmakers display a pattern of fluctuation rather than regularity over time with regard to employment. There is interannual turnover with regard to employers in the brick industry, as well as a pattern of circulation between three branches: brickmaking, construction, and agriculture. Also, in recent years, given the overall decline in brick production and the general crisis of labor in Mexico reflected in unemployment and eroding real wages, it is common for *ladrilleros* to moonlight or to supplement income from brickmaking with income from other part–time, casual jobs. Boredom is one reason often cited for periodically withdrawing from brickmaking and entering construction or agriculture. Economic hardship is the main reason cited for seeking supplementary income from odd jobs. Under conditions of survival in the urban and periurban economy of Reynosa and other border cities, which generate a household strategy of multi–occupationality and of participation in a diverse range of cash–earning activities, there is necessarily a market–driven fluctuation over time in which activities or occupations are practiced and in the degree of participation in them.

Despite the turnover and the possible subordination or displacement of brickmaking by other jobs or activities, experienced molders tend not to permanently abandon this branch, not only because jobs are nearly always available in the unmechanized brickworks producing for the domestic construction market, but also because they can earn a higher daily wage for a shorter, self–directed workday in the brick industry than they can in construction or agriculture for a longer, supervised workday. For example, during 1993 a brickmaker could earn thirty to forty–five new pesos [$9.00 to $13.50] per thousand brick or 120 to 180 new pesos [$36 to $44] for a four thousand brick week of thirty man–hours compared to a day wage of twenty to twenty–five new pesos [$7.50–$10.50] in construction or agriculture for a five–day, forty–hour week under closer supervision.

Cross–Border Trucking: 1990s to the Present

Tiburcio Ramírez began hauling brick from Matamoros to Houston in 1965 and from Reynosa to Houston in 1973. He owns four trailers and, during the boom days, his trucks made a average of twelve round trips weekly between Houston and the border. His trucks transported bundles of used cardboard, used car engines, and cement brick to the border and transported handmade Mexican brick on the return trip to Houston. In the 1990s, he averaged only one round–trip per week, and sometimes one or two per month. He and his wife raised ten children, only two of whom remained in his household as dependents in 1993.

According to him, reduced household expenses enable him to survive on short–haul transport, in the Houston area, mostly deliveries for building supply businesses, supplemented with weekly long–haul trips to the border. He calculates that he cleared about $750 on a round–trip that included bringing a load of brick from the border to Houston. His earnings are derived from transport charges, not from the cargo. He was not optimistic about the future of independent trucking, mainly because he viewed the border trucking business as being heavily penetrated by narco-traffickers or other criminals who he said are not interested in earning much in trucking but, rather, in money laundering.

Ramírez explains border crossing procedures from Mexico to Texas with a load of brick as follows: "From the Mexican side you have an invoice and a customs broker prepares the export papers. When I cross to the other side I have the invoice from the brick plant and I pay five dollars for the import papers. No other taxes or fees are paid. You get a receipt and head for the inspectors. They inspect the trailer to make certain that you are not carrying anything you are not supposed to."

Trailers are not weighed at the border crossing but at weigh stations on the highway. Truck drivers know that they cannot exceed a weight limit of 80,000 pounds including the weight of the trailer. Tiburcio's truck can haul up to twenty-two tons but he usually does not exceed twenty tons—equivalent to fourteen thousand to fifteen thousand eight–inch size brick or eight thousand to eleven thousand nine–inch size brick. Load capacities vary according to the weight of the trailer bed. The day I interviewed him at Reynoso's plant, he was returning to Houston with a load of thirteen thousand three hundred brick.

Most brick haulers like Tiburcio either unload brick at a brickyard owned by a distributor or, at times, directly to a building site. Brick haulers, much to their frustration, are sometimes caught in difficulties between retail sellers for whom they work and home buyers or, more precisely, in disagreements regarding the color or appearance of the brick being delivered to a particular building site. Tiburcio anecdotally recounted one such incident that convinced him that when selling brick to homeowners the retailer should seek to have a consensus among the buyers regarding the type and color of the brick being purchased. He arrived at a building site with a load of palleted brick purchased from him by a Houston builder. However, upon seeing the brick the builder's client refused to accept the brick insisting that they were not the ones he and his wife had selected. Tiburcio spent an hour or so discussing the matter with the husband but then decided to resell the brick to another builder. He did so and had off–loaded several pallets at this second builder's site when he received a call from his original client informing him that the homeowner–to–be's wife had decided that she liked the brick and wanted them delivered again. So Tiburcio reloaded the pallets on the trailer and returned to the original site. Much to his chagrin, when he returned to the original site, the husband and wife were both present and told him that they had decided against his brick. As this anecdote illustrates, until a load of brick is delivered and accepted at the building site, it may not actually represent a final sale.

The complex interplay between nativity, citizenship, residency, and business in the border brick industry today are illustrated in the case of Adalberto [Beto] López [author interview, May 1999] who, since the late 1970s has been involved in trucking brick from the Mexican *ladrilleras* to Houston, Corpus, and other Texas cities. He was born in the state of Nuevo León, Mexico, came to Reynosa in 1952, working first as a tractor–trailer driver for Mexican companies, then as an independent owner–driver. He started hauling brick in 1974 and, at the peak of his business in the late 1970s and early 1980s, had a fleet of six tractor–trailers. He hauled almost exclusively for a large Houston dealer who paid him four dollars per thousand brick hauled. In those years his trucks made an average of twenty–five round trips weekly between the Reynosa *ladrilleras* and Houston. In the decade of the 1990s his six–truck fleet has been reduced to two as he lost four trucks, which he was buying on credit, in the wake of the 1982 economic crisis and crash in the Texas construction industry. He continues to haul for the same Houston dealer but on a much–reduced schedule averaging three or four round–trips per week. In recent years he has supplemented his income by doing other hauling and, also, by selling Mexican brick in the Rio Grande Valley at the site of his trucking business that also has a used truck lot.

Since 1978 Beto has lived and had his business located in McAllen, Texas. He remains a Mexican citizen with a U.S. residency visa but has filed for U.S. citizenship. When I pressed him on the matter of citizenship and national identity, however, his response seemed to contradict, or at least was ambivalent about, his satisfaction with U.S. residency and desire for U.S. citizenship. He emphasized that since the Mexican legislature had recently passed a dual citizenship act, Mexico would continue to be his country [*tierra*] even after he becomes a U.S. citizen. Yet in response to another question I posed regarding the special status and exigencies of being a borderlander and doing business on the border, Beto's response swung, pendulm–like, back toward the U.S. side of the equation. He unfavorably compared the anxieties and frustrations of crossing the border with trucks to do business in Mexico with the rewards and relative ease of making a living in the trucking business in the United States; and also emphasized that he has enjoyed living in the U.S. for the past twenty years.

The ambivalence so palpably expressed by Beto López, epitomizes the dilemma of untold numbers of hardworking and enterprising lower Rio Grande border people whose social and economic interests, for generations, have compelled them to be peripatetic border crossers. Because the borderline they cross divides two nation–states, these borderlanders are inevitably confronted throughout their adult lives with the need to reconcile conflicts between making and earning a living, residence, nativity, citizenship and, ultimately, to negotiate their very identity. The extent to which many families like the Sotos and Lópezes have risen to these challenges and, over the years, have successfully negotiated the border labyrinth to their own advantage is worthy of admiration. This issue will be examined further in the next chapter through the case of Tony Reynoso.

One of the Reynoso's two tractor trailers is licensed for hauling brick from the Reynosa plant to the yard in Pharr, Texas but also to deliver brick within the Rio Grande Valley. The second truck is licensed to transport brick loaded in Pharr to any off–border destination in Texas. In January 2008, this truck was making several round trips weekly to distributors in Corpus Christi and Houston.

In 1995 the Pharr–Reynosa International Bridge was opened to handle the increased commercial traffic generated by Reynosa's growing *maquiladora* or free trade zone, mainly involving containerized cargo of electronics components or products. Its opening was coordinated with the closing of the older Hidalgo–Reynosa Bridge to commercial traffic. The Pharr–Reynosa Bridge is touted for its state–of–the–art technology, implemented by both U.S. and Mexican Customs, incorporating laser and gamma ray equipment to expedite cargo inspection and reduce crossing time. This system may function smoothly for the high volume carriers of containerized cargo from the free trade zone but has not done so for non–containerized cargo like brick loaded on flatbed trailers. The Reynosos' experience has been that about half of the four or five crossings of their truck per week results in full inspections requiring each pack of brick to be removed from the truck for separate electronic inspection. Sometimes the strapped shipping packs are opened and brick removed. The cost of this procedure is one hundred dollars per occurrence payable at the time of inspection by the carrier. In 2006 the Reynosos estimate that they paid an average of two hundred dollars weekly for inspections simply because their cargo did not conform to the norm for the Pharr–Reynosa Bridge. Moreover, despite its claim of expedited crossings, the Reynosos found that bridge to suffer from back–ups and delays of eight or more hours to await inspection, especially on weekends.

Negative experiences at the Pharr–Reynosa bridge have caused the Reynosos to consider alternatives. They know that the Camargo–Rio Grande City Bridge has a long history as a crossing point for clay products manufactured in Camargo, as well as in the Monterrey, Nuevo León and Saltillo, Coahuila areas. Although this route is about sixty miles longer and costs them a thirty dollars per trip surcharge for extra diesel fuel, the inspection procedure at this bridge usually proceeds normally and without time–consuming and costly unloading and reloading. The end result is that the Reynosos save money by using the Camargo–Rio Grande City bridge for most of their truck crossings. Eventually, the planned new Anzaldúas–Mission bridge may well prove to be the best crossing option for the Reynosos. Indeed, they have bought land in the Mission area to eventually relocate their Pharr office and yard near the site of the proposed bridge.

Customs Issues and the Pallet Problem

Beto Lopez' remark that he is always apprehensive when his trucks pull out of his McAllen yard to cross the international bridge to Reynosa until they return is symptomatic of a history of misfortunes and frustrations experienced over the decades with the customs bureaucracy, hijackers, or other sources in moving trucks and/or cargo to "the other side." My interview files are replete with accounts by brick exporters, importers, and truckers alike of many incidents of capricious or arbitrary harassment or enforcement by Mexican customs officials when trucks enter Mexico unloaded or return from Mexico to the U.S. loaded with brick. Over the years tractor–trailers with U.S. registration have reportedly been impounded, stolen, or damaged in Mexico; and *mordidas* [bribes] regularly extorted by Mexican customs officials of loaded trucks returning to the U.S. from Mexico, regardless of whether the trucks' country of registration or the status of the customs documentation carried by the trucker.

The hope of everyone concerned with the brick trade was that the implementation of NAFTA [North American Free Trade Agreement] would bring about permanent changes to facilitate commercial border crossings—especially truck transport. Beto Lopez's testimony suggests that these expectations have yet to be met. He complains of erratically changing bureaucratic regulations over the years—including (1) the *Mexican Federal Vehicle Registry* requiring Mexican permits and plates for American trucks to enter Mexico [to which he attributes the theft of one of his trucks in Mexico], (2) demands for a special permit for the transport of export merchandise from Mexico to the U.S. which has cost him as much as three thousand dollars to get one truck in "noncompliance" released from impoundment by Mexican authorities, and (3) refusal of entry into Mexico for lack of the required permit of the day. In frustration he noted: "Many times I have not been able to take my truck into Mexico because, for one pretext or other, Mexican customs won't let me pass. This depends on the whims of officials who happen to be there at the time, asking for unnecessary documentation or something like that."

In May 1999 the border–crossing crisis of the moment involved wooden pallets on which brick are stacked, wrapped in plastic, and loaded in the Mexican plants for transport to the U.S. Without prior warning Mexican customs began refusing to allow brick hauling tractor–trailers to re–enter Mexico carrying empty pallets. After several inquiries by irate Mexican brick exporters with Mexican customs, it was determined that the problem could be solved, at least temporarily, by typing "The pallets are returnable" on the export permit for each load of brick.

Joel Rodríguez was at a loss to understand why Mexican Customs suddenly decided to make an issue of pallets which had been transported back and forth across the border by brick haulers for more than ten years without incident. In his opinion, even if the Mexican authorities were simply enforcing a law that had been on the books but never enforced, the law was a bad one which represented government harassment rather than facilitation of bilateral commerce. Rodríguez also complained that his is still a Third World industry that cannot afford to pay customs agents a one hundred dollar fee to process the paperwork required to export eighteen to twenty pallets worth one dollar each. In short, the pallet problem is simply the most recent episode in a decades–long history of harassment and frustration in trucking brick across the border.

In another twist on the *maquiladora* relationship, smaller–scale *ladrilleras* like Toño Reynoso's were unaffected by the problem of pallet importation simply because they acquired used pallets from the nearby *maquiladoras* and sell them with each load of brick. They could arrange with truckers to have the pallets returned but would have to pay a fee for that service since the truckers main business is with the larger, gas–firing ladrilleras. In short, it is less costly for the small–scale *ladrilleras* to buy pallets from the *maquiladoras* rather than attempting to reuse them. Incidentally, that connection also proves advantageous in that the small–scale plant operators buy broken or discarded pallets from the *maquiladoras* to use as fuel in firing their kilns. This activity also helps the wood–burning plants to keep costs and prices of fired brick below those of the gas–firing plants.

Conclusions

Staying in business under conditions of an unpredictably fluctuating market and a dwindling supply of labor, in a labor–intensive industry, requires an appropriate strategy. Labor must be retained and, therefore, production must continue whether or not orders are on hand. Hence, inventories are built up to await market upturns. When a market upturn is prolonged and inventories are depleted then pressures build to expand production. This is what appeared to be occurring in 1999.

An increase in demand for Mexican handmade brick in the early months of 1999 was substantial enough to reinitiate, albeit on a much reduced scale, the satellite field strategy in one of the two gas–firing plants, as well as intensify competitiveness among plant operators for available brickmakers, especially the small–scale plants which are more prone to labor turnover. The smallest plants, with only unsure access to casual laborers and a minimal pool of dependent household labor, are the most economically squeezed units in all market conditions. Tano Arriaga's plant, due to a scarcity of labor and cash, agreed to sell its entire output of crude brick to a gas–firing plant for a price of forty–five dollars per thousand. It is more favorable for Arriaga to sell unfired brick rather than to fire his own brick and then wait for buyers to show up who, in any case, probably will not pay more than sixty to sixty–five dollars per thousand for fired brick.

Cross–border transport might appear to be the least problematic link in the chain of processes that enable Mexican–made brick to be mortared into the walls of Texas buildings. Such an appearance would be misleading, however, simply because it is the one link in the process that is most vulnerable to binational politico–administrative/bureaucratic rules, regulations and interventions, with the latter sometimes based upon arbitrary judgments and/or thinly veiled corruption. Truckers, in effect, run a sort of gauntlet during every border crossing, whether their truck is loaded with cargo or empty, and regardless of the direction it is headed or its country of registration. This is true simply because of the licensing, taxing, and inspection regimes truckers with their trucks, and/or the businesses that employ their services, brick manufacturers or distributors must negotiate every time an international bridge is crossed. Fortunately, since brick are not perishable merchandise, border–crossing delays result only in losses of time and money, and are routinely absorbed by the brick industry as costs of conducting business as usual.

Notes

1. The common use of the Spanish verbs *echar* [to throw or toss] and *cortar* [to cut] by the molders with reference to the molding process, as illustrated in the case of Jose Angel Agüero, gives rise to two names also used to refer to the one who performs this task, namely, *echurero* [the thrower or tosser] and *cortador* [cutter]. In my 1998 book I used the term *hechurero*, rather than *echurero*, as a cover term for brickmaker or brick-molder on the assumption that it was a derivative of the Spanish verb *hacer* [to make] and therefore implied *ladrillos hechos por* [brick made by] persons so designated. Since in spoken Spanish the initial 'h' in words like *hechura* or *hechurero* is silent there is definite ambiguity in everyday usage among brick industry workers regarding these two words and *echura* and *echurero* since they have the same pronunciation. In the Agüero case, his reference to brickmolding as *echar ladrillo* seems to indicate that he could correctly refer to himself and others who mold brick as either *echurero* or *hechurero*.

My copy of the *Diccionario de la Lengua Española of the Real Academia Española* (Madrid 1992, p. 1090) has the following definition for *hechura*: "*Acción y efecto de hacer* [Action and effect of making] that is perfectly applicable to the process of molding brick. The term *hechurero*, although not in this dictionary, is understood by Spanish-speakers to mean "one who makes" or "maker" and, when used in the context of the brick industry, would imply making or molding brick. The same dictionary also has an entry for *echura* (p. 789) that is defined as *Echada o tiro*, that is, something thrown or tossed. Again, the term *echurero* [not in the dictionary] would appear to be a legitimate derivative of *echura* to refer to a brick molder in the sense of *thrower* or *tosser* of clay into the mold.

Another way brick industry workers, like Agüero, refer to molding [or at least a specific step in the process of filling a mold] is *cortar*, so the person who performs it is known as a *cortador*, implying that a portion of wet clay is *cut* from the main mound to provde just enough to fill two compartments in a six compartment mold [three such cuts of clay being required to fill the mold]. I conclude that in the border brick industry every *molder* [*moldeador*] is a *maker* [*hechurero*] is a *cutter* [*cortador*] and is a *thrower* [*echurero*].

Just for the record, when I discussed this semantic issue with Tony Reynoso, he admitted that ambiguity is present in the Spanish terminology, even for native Spanish-speakers, but insisted that the correct usage to identify a brickmolder derives from the Spanish verb *hacer* [to make] and is, therefore, *hechurero*.

Chapter Nine
The Present and Future of Handmade Brick for Texas

Bordering the main unpaved road running north–south into the middle of the northwestern section of Los Longoria, at its junction with another unpaved road known locally as the *Camino Real*, lie the fallow fields, and the long–abandoned buildings and irrigation canals of a once flourishing *hacienda* belonging to the Longoria family. Now overgrown with grass and brush, the *ex–hacienda*, which originally encompassed several hundred hectares, was reduced to a few tens of hectares by a 1941 agrarian reform decree.

At this same *Camino Real* junction looking to the northeast, by far the most prominent features in the landscape are an imposing gated and fenced expanse of land with stacks and stacks of palletized brick and an imposing brick kiln in the foreground, and an elongated roofed structure in the background. This multi–hectare brick plant, bustling with activity of men and machines, belongs to a non–ranching family of craft industrialists, the Reynosos. The juxtaposition of the quiescent ex–*hacienda* of the Longoria family, well ensconced in the ranks of the traditional Reynosa elite, with the bustling *ladrillera* of the Reynoso family, migrants to the border from Guanajuato, via Zacatecas and Coahuila where they learned brickmaking, is a metaphor for the economic transformation of Reynosa over the past half century [See Appendix 1. Photos 63 and 64 provide panoramic views of the approach to the Reynoso plant].

Since the Reynosos became residents of Los Longoria after the 1941 agrarian reform and, in any case, were not interested in agriculture or ranching, they could not qualify as *ejidatarios* but were able to acquire land suitable for brickmaking from private landowners and *ejidatarios*. As explained in chapter 3, Diego Reynoso and his stepson, Sebastián, were not natives of Reynosa and did not move to Los Longoria to take up brickmaking until the early 1950s. Antonio 'Toño' Reynoso took over day–to–day management of Sebastian's brick plants in the early 1980s, and in the 1990s purchased Sebastian's interest and an additional hectare of land adjacent to the old plant site. In the late 1990s, Toño was fully engaged in a project to expand and upgrade his plant.

Unfortunately, this project was interrupted by Toño's serious injury in a traffic accident in 2000. As it turned out, however, this interruption was short–lasting and, in fact, proved to be a stimulus to the business due to the entry of a new partner at *Ladrillera Reynoso* in 2001, namely Toño's son, Tony. Thanks to a new infusion of capital accumulated during Tony's stint in a Dallas construction business the plant's redevelopment plan was accelerated. After several years of joint effort by father and son, an entirely new plant has been built that in terms of appropriate technological innovation and entrepreneurial orientation is unprecedented in the history of the Reynosa brick industry.

Transforming *Ladrillera Reynoso* for the Twenty–first Century Market

The new Reynoso plant's identifying bilingual slogan "The old handmade brick with a contemporary vision" [*El viejo ladrillo hecho a mano con una visión contemporánea*] provides a hint of widespread recent innovations introduced to achieve the Reynosos' goals of increasing the quantity and variety of brick produced—in terms of types, sizes, and colors; and to improve its quality by lessening percentage of water absorption and increasing compressive strength to levels that comply with ASTM [American Society for the Testing of Materials] standards. Indeed, the slogan understates the emergence of a new Reynoso plant along the unpaved *Camino Real* adjacent to the old plant and to the remains of the defunct Longoria *hacienda*.

Beginning in 2001, a series of investments were made precisely in those stages of the production process where labor–saving mechanization is most effective without changing the authentic handmade quality of the brick: the extraction, transport, and grinding of the clay, its mixing into a soft–mud, and the handling of brick prior to and after firing. The stimulus for these innovations came from the emergence of a partnership between Toño and his son Tony, following a serious highway accident in which Toño was badly injured. At the time of the accident, Tony was working in the Dallas construction industry where he went to seek his fortune and improve his English following his graduation in the mid–1990s with a degree in architecture from the Autonomous University of Nuevo León in Monterrey. Although he grew up as a worker in his father's brick plant, and all of his pre–university schooling was in Reynosa, his father sent him to live and work with relatives who had a carpentry business in the Rio Grande Valley during the summer and school vacation periods so that he could learn another craft and experience life north of the border.

After a rocky start in the Dallas construction industry that included being cheated out of twenty thousand dollars in salary by an unscrupulous boss, being broke, and sleeping for a time in parked trailer cabs with a friend's permission, Tony managed to survive and eventually flourished in the same industrial niche where he had been cheated, namely, completing carpentry jobs in large projects that had been unfinished or done improperly and were, therefore, on a punch list requiring completion prior to closure. His successful stint as boss of a team of three groups of workers motivated him to start his own subcontracting company that grew to two hundred twenty–five employees, and earned him hundreds of thousands of dollars in profit in the space of a few years. To make a long story short, in 2001 Tony paid his father $200,000 in cash for half–ownership in the brick plant, and immediately started investing more of his Dallas earnings by buying two hectares of land from the Longorias located next to the one hectare parcel where his father had relocated his plant a few years earlier. New machinery and equipment was acquired, and new kilns and facilities were constructed.

One of the main reasons for this investment was a perceived need to expand output. In Tony's words: "If I was going to work at the *ladrillera* I had to double output so that two families [his father's and his own] could live from the same business." He also recognized that doubling the plant's output would carry with it the obligation of a more comprehensive and aggressive approach to sales that had never been of concern to his father. Tony commuted to Dallas on a regular basis to attend to his construction company's gradually downsized, but still profitable, operations until he decided to commit full–time to the brick business in late 2004.

The focus of the Reynosos' investment plan, derived from their deep understanding of the process of making handmade brick, was to improve the work conditions and productivity of the *ladrilleros* who Tony poetically referred to as "people with their feet in the mud" without whom there would be no handmade brick ["*Cuando digo 'ladrilleros' me refiero a la gente con los pies en el lodo, que es la mas importante en el ladrillo hecho a mano*"]. Tony expressed this point more completely as follows:

> To increase output in a handmade brick plant is complicated. Within a long process of fabrication, there are some areas where machinery can be used to accelerate it such as extraction [of the raw material], transportation, milling of the clay, mixing of the mud, and transport of unfired and fired brick. But the principal activity, the one that consumes the most time in the fabrication process and that has to be done by hand, is molding [*moldeo*] or making [*hechura*] and demolding [*volteada*] on a dirt floor.

In other words, the Reynosos realized that to increase production at their plant they had two options: they had to contract more molders, or they had to help the molders perform certain tasks more efficiently so that the productivity of each worker could be increased. The first option, under prevailing labor market conditions, would have been problematic since experienced brickmolders are hard to find, and the training of inexperienced workers to mold is time consuming and has a low success rate—out of ten workers given tryouts only two will stay at the job. So, the second option was preferable, and the most strategic labor–saving innovation was clearly the mixing of clay.

Changes made at the plant since 2001 as a result of Tony's investment can be outlined under the following headings: 1) clay extraction; 2) clay storage; 3) clay preparation; 4) clay molding and drying; 5) transport of green brick to kiln; 6) kilns and firing; 7) packing, and, finally, 8) inventory, product lines, and marketing. The reader should keep in mind that even after Toño relocated and upgraded his old plant during the 1990s by purchasing new equipment and rebuilding two kilns, clay preparation remained unmechanized and brick molding and drying was still conducted in the open air and under full sun, unprotected by a roof. If the reader can locate the Los Longoria–Calabazas landform on *Google Earth* and zooms–in on its western side, the new Reynoso plant site cannot be missed since *it is marked by the largest visible structure [the molding and drying shed].*

Clay Mining or "Winning"

With the use of backhoes and dump trucks, several of which have been added to the plant's inventory, clay is extracted from the sides of the pit with straighter cuts and is transported faster to the main plant than it was previously when clay was dug by hand with shovels. Another advantage of the more mechanized winning of clay is that after it is extracted the land can be re–contoured so that it blends in more with the natural landscape. Such conservation measures were too time–consuming and costly to undertake previously.

The lower section of the Reynoso plant site where clay is extracted, encompasses about four hectares. It has different veins of soil deposits that, upon visual inspection of color and texture, are determined to have relatively high or low proportions of sand to clay content. The preferred type of clay has a higher content of iron oxide since experience has confirmed that brick made from this type of clay are easier to fire, harder, more durable, and when fired have a highly marketable pinkish tint. However, the daily challenge at the clay pits [*pozos*] is to try to make sure that each truckload of raw material is evenly divided between clay and sand content. Too much clay content in the mud produced by the mixing process will result in brick that tend to crack or split [*se quiebran*] during firing. On the other hand, too much sand content results in brick that tend to crumble or fall apart [*se desmoronan*]. Over the past twenty-five years about one and one–half hectares of soil have been mined so the Reynosos estimate that they have, at least, another twenty-five years to mine the remaining deposits on land under their ownership. However, land contiguous to theirs is available for purchase or rent if the need to do so arises in the future. In short, the supply of their principal non–renewable raw material is elastic. [See Appendix 1. Photo 65 shows a clay pit at the Reynoso site].

Clay Storage

A shed was constructed with a cement floor, cement block walls and a corrugated roof to store clay in anticipation of rainy weather that makes extraction from the pit impossible. The front of the shed is equipped with a curtain that permits clay to protect the stored clay from getting wet from blowing rain, and the floor has a hopper that can feed the stored clay onto a conveyer belt that carries it to the nearby mill and sifter. Wet clay cannot be processed since it jams the mill. The current storage facility has enough space to hold only a one–day supply of dry clay. If rainy weather lasts longer the mill is shutdown and wet raw clay is trucked directly to the molding patios.

Clay Preparation

The new mill was built with an integrated sifter to eliminate pebbles and other small foreign materials called *coco* from the clay to give it a finer and more uniform consistency. This facilitates the mixing of the dry clay [and some sand] with water to form the muddy mixture [referred to interchangeably as *lodo*, *batido* or *soquete*] that is used to mold brick. The advantage of a fine, uniform clay mix is that it diminishes porosity in the brick molded from it, thus making them more resistant and less absorptive after being fired.

The Reynosos purchased a used screw–type clay mixer [*batidora*] from Efraín Salinas and rebuilt it. After the raw clay is conveyed from the storage area to the mill, the milled material is conveyed into a large hopper that drops it onto another conveyor belt that feeds upward into the mixer where the milled material is mixed with water. After mixed to a desired consistency with approximately 30 percent water content, the mud is fed by gravity into a dump truck parked below the mixing station. One truckload of raw clay, enough for molding two thousand brick, can be processed into mud in one hour from the time of its arrival at the mill. As a general rule, six truckloads of raw clay, enough for twelve thousand to thirteen thousand brick, are processed daily.

The mill and mixer operator has to pay careful attention to the consistency or degree of wetness of clay because truckloads of raw clay that arrive for processing do not have uniform moisture content. It is up to the operator to judge how much water must be added to the milled clay after it is conveyed to the mixer to give it the required consistency. He is able to add water directly into the mix from a hose, connected to a well–fed tank, that he controls at the mixing station.

The principal advantage of the mechanical mixer is that it enables molders to spend more time molding since they are no longer responsible for preparing their mud supply from scratch as they were previously. According to Tony Reynoso, with the combination of mechanical mill and mixer, "Mud is mixed better, more uniform and not as much foreign material such as pebbles as before. The outcome is a better clay mix. The principal thing is that a molder can mold fifty percent more brick with the machine–mixed mud than he could when he had to mix his own mud."

The normal amount for a molder who also prepares his own clay batch is one thousand brick daily. With the machine–mixed mud he can now mold fifteen hundred in the same amount of time. From the Reynoso's perspective this meets their goal of expanding output and they do so, in operational cost terms, by employing only two additional workers: one who is in charge of the milling and mixing operation and delivers mud to the molder's *patio*, and another who excavates the clay at the pit and delivers it to the mill. The only additional operational cost incurred is for electricity to operate the machinery. Maintenance and upkeep of the machinery is done by this employee or by the Reynosos themselves.

Initially, molders at the plant resisted the changeover to pre–mixed clay since they claimed that it did not have the consistency to which they were accustomed to prepare for themselves. Like bakers who are accustomed to mix their own dough from raw flour, and reject pre–mixed dough, the molders claimed that the pre–mixed clay was either too watery [*aguado*] or not watery enough. My conversations with several of them confirmed that they now have become accustomed to pre–mixed clay, not only because it makes it possible for them to make more brick per day and increase their incomes, but also because they have water available to remix [*rebatir*] it to the consistency they prefer. During three months or so in 2007 when it was not possible to operate the mill due to heavy and prolonged rains that soaked the clay pits and made the clay unfit for mechanical milling, the molders complained about having to mix clay because they had become accustomed to the advantages of pre–mixed clay. In short, the mechanized processing of clay is now fully embraced by the molders. Tony expressed it this way: "Now that they can make more brick and earn more money they no longer resist mechanization."

Although the Reynosos had to invest in the machinery and other equipment necessary to assemble the clay milling and mixing plant, they assembled it themselves through a time–consuming trial–and–error process that lasted three years. Initially, they constructed an enclosed building to house the entire mill but found, much to their chagrin, that there was not sufficient space in the building to properly install the system. So, they converted the original building into a workshop, office, and company store [*tienda de raya*] and proceeded to construct an open, partially roofed building that would better accommodate the two conveyors, the grinder, the mixer, and the dry clay storage shed. In Tony's words:

> What we invested here is a lot of time. In order to know how many blades we had to put on the mixer, it was trial and error. We put less than it had when we bought it and found that the mud would not advance through it. We had to add more. And then we had to put it into a more inclined position. In short, to get where we are now (2007)—that is a six–month period during which the system has not failed—three years passed.

When I raised the issue of energy cost [electricity and diesel fuel] of running the plant by asking Tony's father, Toño, if he had nightmares about his new dependency on costly energy, he laughed and mentioned that Juan Tueme, one of the near legendary personalities among Reynosa's brick plant owners, was an enemy of mechanization and much admired the antiquated system of milling and mixing clay with horses and mules that were fed from the leftovers of corn harvests. There is no doubt that Tueme's view resonates to a certain extent with Toño's. Nevertheless, Toño is now more inclined toward his son's perspective regarding mechanization [See Appendix 1. Photos 66–68 illustrate the mill facilities and activities as described]. Tony summed up their position as follows:

> In reality, we do not see this as an investment from which we have to make a profit but as part of the production process. Let's say it's like buying a dump truck. For us, up to a point, we don't see it as the truck paying for itself. Rather, we see it as part of a process that in the last analysis will increase quality, increase output, and give better conditions for the workers. Like the operator of this mill who, before the roof was put up, had to work in the sun.

Clay Molding and Drying

These processes now occur within the confines of a new shed ['*techo*', '*secadero*', or '*casiller*'] measuring 350 feet in length and seventy-five feet in width, providing ample molding and drying *patios* for ten molders and their output. Each molder is assigned a *patio*, measuring approximately thirty-five feet wide by seventy-five feet long, which has sufficient space under the roof to accommodate the clay supply [*montón de lodo or soquete*], two piles of sand [coarser one for the mold and a finer one for the hands to keep mud from sticking], the floor [*piso*] area to lay out several lines [*lineas*] of newly molded brick and, finally, space to stack drying brick in towers after they are removed from the floor. Each row in a line typically contains a turned–over moldful [*moldera*] of six brick, and the seventy-five foot length of the *patio* provides space for fifty-five to sixty *molderas* per line for a total of 330–360 brick, the actual number varying in accordance with the length of the brick.

Several raised concrete platforms, about four feet wide, have also been constructed to accommodate stacked brick. A roofed drying area prevents brick from drying too rapidly as they do when exposed to direct sunlight which leads to a higher rate of breakage. Given the slower and more uniform drying process with the roof, brick are now more durable. Molders also benefit from the roof—first because they work more comfortably in the shade and second because they no longer have to worry about covering brick with plastic tarps to prevent rain damage.

The mechanization of clay preparation and of its delivery to molding areas translates into more durable brick and higher productivity per molder but does not change the appearance of brick that continue to show the idiosyncrasies of each individual molder. Tony expressed this as follows: "I can show you the difference between one molder and another independently of the fact that the mud they mold is exactly the same. Each molder makes a brick that is unique to him. And even though you may try to make a brick equal to it, you cannot because each set of hands is different."

Tony has learned how to identify brick made by a particular brickmaker at his Reynosa plant in the inventory of fired and packaged brick in his sales yard in Pharr, Texas. I began to understand the empirical basis for his ability to do so during a visit, in his company, to the molding patios at the plant during the week of October 22, 2007 when seven molders were at work daily. Only one of the seven has the knack and willingness to make any one of a dozen basic shapes and sizes of brick produced by the plant. The remainder of the molders tend to specialize in one or two different–sized brick. Individual molders do not put their initials or any other identifying mark on the brick they mold. Each compartment of most wooden molds has attached to it a metal brand plate or [*placa*] that simultaneously creates an indentation [frog] and stamps the ***REYNOSO*** trademark with *Mexico* [country of origin identifier required by U.S. Customs] under it on each brick. Also, all hand molded brick have ground marks on their undersides [the side that is turned out of the mold onto the ground]. So, the question arises, what is the empirical basis for the distinctiveness that sets one molder's brick apart from another's?

One relevant factor is that even when pre–mixed clay is delivered and deposited as a moist mound on the molder's patio, the molder remixes [*rebate*] clay in the process of *cutting* it to form a smaller ball to put into the mold. The careful observer of this process can detect differences between molders in the way it is performed, as reflected in the amount of water added to the ball or the force with which the ball is thrown into the mold, that may translate into personally distinctive brick especially related to smoothness and consistency of texture.

Another factor has to do with how the mold is turned over to deposit the moist brick on the ground and how the mold is pulled off the emptied brick. The key variable here is kinetic, that is, the pace and degree of exertion with which the molder performs the acts of mold turnover and withdrawal. If the turnover is too forceful then when brick hit the ground they tend to twist or become slightly uneven in shape; if the mold withdrawal is too slow or slightly deviates from the perpendicular, the deposited brick tend to have rough edges or corners called "lips" or "lipping" (Gurcke 1987:108). The resulting differences are, admittedly, subtle and to the unpracticed eye nearly invisible.[1]

Most importantly in terms of the aesthetics of the brick market these minor differences in degree between experienced molders are inconsequential in the context of comparing either machine–molded or extruded brick with hand–molded brick. The aesthetic appeal of the latter lies precisely in their visible flaws or imperfections, their minor deviations from linear straightness, uniformity in surface texture, or consistency in color. In the well–chosen words of a German expatriate art historian who lived and worked in Mexico, Alexander Von Wuthenau: "The grace of imperfection is worth more than graceless perfection" (Rosa Pérez, personal communication; see Kelly and Street–Porter 1989:7).

Over the past three years the plant has experienced turnover in its staff of molders. From a total of eight employed during 2005, the staff dwindled to five in 2006, but rose again in 2007. When I visited the plant in October 2007, seven molders were at work, only three of whom—Pedro Escamilla, Juan Garza, and José Luna—were on the staff in 2005. One of the newcomers, Manuel Pasilla, originally from Durango, was employed for several decades in a Camargo plant that shutdown in early 2007. Despite his advanced age of seventy-three, he came to Reynosa to look for work and has molded an average of 4,500 brick per week at the Reynoso plant since being hired in the summer of 2007. Another molder, José Luna, who has worked for the Reynosos for many years is also in his seventies and during September 2007 molded 21,850 pieces for a weekly average of 5460. During the same month, Juan Garza, who was sixty-seven years old, molded a total of 27,710 brick for an average approaching seven thousand pieces per week. Garza out-produced younger molders, during that same period, by two thousand to seven thousand brick. Plant records covering thirty-four weeks in 2005, 2006, and 2007 indicate a weekly average output per molder of about four thousand pieces. That number is exceeded by the 5,650 average of the three oldest molders Pasilla, Luna, and Juan Garza during the month of September 2007. The attitude of these senior molders is that they will keep working as long as they are physically able to do so. Juan Garza spoke for all of them when he told me in a jocular way that if I showed up at the plant sometime in the future, and did not see him at work, the reason would be that he had died on the job.

The average age of forty-five of the seven molders at work during my October 2007 visit to the plant, points to a possible future output crisis. The youngest molders are in their thirties, the oldest in their seventies, and there are none in their teens and twenties. Humberto Garza, one of the younger molders, produces a respectable weekly average of six thousand pieces but is not enthusiastic about the job which he considers to be too hard on his knees. He prefers setting brick in the kiln and is simply biding his time as a molder until there is an opening as a setter. He worked in past years at the old plant as a setter but did not adapt well to the changes introduced after 2001, and was switched to molding. He quit and spent a few years elsewhere before returning to the plant in 2007. If he is unable to return at some point to his preferred job as a setter, he is likely to quit again and seek alternative employment—probably in the *maquiladora* or construction sectors for less money than he makes as a molder. [See Appendix 1. Photos 69–82 show the new roofed molding/drying space and two different *hechureros* at *Ladrillera Reynoso* in 2007—Juan Garza and Manuel Pasilla—performing the tasks described above].

Transport of Crude Brick to Kilns

A new heavy duty forklift, twenty specially constructed loading containers with 1,000 brick capacity each, and a network of concrete roadways have radically improved the plant's system of transporting crude brick from the storage shed directly to the kilns for loading. Previously, this process was accomplished with carts manually pulled across the bare ground, and brick were set in the kiln by manually taking them from the carts and tossing them in relays to a particular position in the kiln. Now crude brick are manually loaded into the boxes [metal frame with wooden sides], then moved by forklift to the setting position inside the kiln, either at ground level or higher up, to facilitate manual removal and placement of the individual brick. This system was initially resisted by setters who were accustomed to the manual relay method of setting but they were quick to adapt and to appreciate its advantages. The combination of loading containers and the paved roadways also reduced breakage of crude brick in the process of moving them from the drying area into the kiln. [See Appendix 1. Photo 88 illustrates the setting of green brick into a kiln].

Kilns and Firing

Since the goal of the new plant was to double the output of the old plant, the Reynosos had to expand firing capacity by building two new kilns. The old plant had two periodic kilns, built about fifteen to twenty years ago by Toño and still in service, with capacities of 110,000 brick each. Since 2001 two new periodic kilns were constructed, mostly of mud and old brick, with capacities of 150,000 and 85,000 brick each. A thin outer layer of cement covers the exterior mud and brick walls of the new kilns to avoid rain damage.

Unlike the older kilns, the two new kilns although still updraft are of the vault [*bóveda*] type with interior arches and closed construction. Their advantages are that they make unnecessary the manual sealing of the top of the kiln [*tapada*] prior to firing, require lower maintenance due to lesser susceptibility to cracking or splitting open, have better heat retention and more internal temperature uniformity, and due to reduced air flow they retain smoke longer within the kiln with the effect of reducing contaminating emissions.

The largest of these new kilns is open at both ends so as to provide more flexibility in setting and unloading. They are constructed with mud and old brick with the walls ranging in thickness from three to eight feet—the thickest and heaviest being located at those points on both side walls, at approximately a height of ten feet, where the ceiling arch and heat intensity is most pronounced. The increased thickness of walls at these points keeps them from fracturing or splitting open from the intense heat in which case the ceiling would collapse. The Reynosos, unfortunately, learned this by trial and error since they were forced to tear down and rebuild the largest kiln precisely because its walls cracked badly after a test firing. This required an additional year's labor for four full–time workers. The tops of the ceilings in these kilns have the thinnest walls and have a few small round escape or chimney holes at intervals running lengthwise.

Regarding fuel to fire kilns, in the 1990s cut mesquite was replaced by scrap wood. This is procured from businesses in Reynosa's free trade zone that rebuild pallets for the *maquiladora* industry, since U.S. law prohibits the re–importation of used pallets. The Reynosos park empty dump trucks at these businesses, and when the trucks are filled with scrap wood they are driven to the brick plant, emptied and then returned to the pallet businesses to be refilled. Extension of the natural gas line installation to the new Reynoso plant is prohibitively expensive since *PEMEX* will not subsidize the cost of installing the delivery infrastructure.

In 2005 Tony visited a factory in Spokane, Washington that produces burners that burn waste vegetable oil. He bought one and has experimented with it since then with mixed results. It is rather complicated to use and maintain but he has also introduced modifications that he has observed in use in Guanajuato, Mexico among potters employing ingenious combinations of motors and blowers that are relatively cheap and easier to maintain. To date he has been able to fire kilns with about a fifty–fifty mix of recycled oil and wood. However, his goal is to achieve greater dependency on recycled oil that is collected periodically in steel drums from various restaurants in Reynosa. The oil from these drums is transferred to a tank on a elevated platform and is gravity fed into the burner and oil delivery system that consists of a series of pipes introduced into the kilns through several fuel ports [*bocas*] on the sides of the kilns. When wood is used as fuel it is also introduced into the kilns through these fuel ports. During my visit to the plant in January 2008, the two fuels were being used in combination: one side of a kiln was being fired with the waste oil system, and the other side with recycled pallets. The two fuels yield comparable results, although their different combustion patterns require careful monitoring.

As a rule of thumb in the brick industry, color is a function of the mineralogy of particular clays, kiln temperature and atmosphere and, to a lesser degree, kiln fuel. Natural Reynosa clay typically has varying amounts of impurities including sufficient iron oxide to make it "pink–burning" under temperatures around six hundred degrees Centigrade (1,112 degrees Fahrenheit) where it achieves hardening or vitrification. At higher temperatures when gas–fired this clay tends to burn uniformly buff or yellowish in color. The combination of the available clay and the relatively humid weather limit the average length of kiln firing in Reynosa to a minimum of seven days regardless of the type of fuel used. A complete cool–down takes about the same length of time. Thus the complete firing and cool–down cycle takes about two weeks.[2]

The Reynosos' experimentation with kiln firings using wood and oil in combination has shown that particular temperatures are easier to maintain with recycled oil but that wood lends itself more readily to changing temperatures beyond the threshold of six hundred degrees Centigrade in order to achieve certain desired effects in the brick being fired—especially brighter colors or mixed color effects, including black smudging, that are produced at higher temperatures and with more smoke. In short, there will always be some production advantages derived from firing with wood but, once the bugs are worked out of the oil burning system, the intention of the Reynosos is to rely on it for 80 to 90 percent of their fuel needs.

The process of building new kilns and of trying out a new fuel has meant that old personnel have had to acquire new knowledge through trial and error and that new, younger personnel have had to be trained. In the last five years two new stokers have been hired and trained out of more than twenty who were given trial runs. The job of stoker is the most difficult job to fill in the handmade brick industry, even moreso than molding, simply because it requires a special tolerance for heat, the risk involved in firing combustible fuels, and for handling the burden of responsibility for tens of thousands of brick that can be damaged or ruined through mis–firing. Stokers experienced with wood burning tend not to want to switch to another type of fuel so the new stokers were trained to specialize in oil burning. Since years of experience are required to develop a second sense regarding kiln temperatures, as well as how and when to stoke the fire, the Reynosos bought a pyrometer so that the newly trained stokers could take exact temperatures during firing. Veteran stokers, accustomed to wood or natural gas firing, can accurately judge internal kiln temperatures by sight or by feeling the heat generated by the kiln from a few feet away. [See Appendix 1. Photos 84–87 and 89 illustrate kiln construction and various features of kilns related to setting of brick and firing the kilns].

Packing of Fired Brick

Another technological innovation in the new Reynoso plant was the purchase of equipment to strap and wrap plastic around brick of the same size to form cubes. Before, brick were simply stacked loose on pallets and transported without strapping or wrapping. The new system facilitates counting and handling brick at the yard, better weight distribution on the flatbed trailers that haul brick, and promotes more convenient handling of brick at construction sites.

Inventory, Product Lines, and Marketing

Although the goal of doubling output through the process of mechanizing and building a new plant was reached earlier, it was not until the start of 2006 that the Reynosos had built up inventory to a point where they felt they could assure their clients fast delivery on orders and where the inventory was diversified to include a wider variety of shapes, sizes, and colors of brick. Once this stage in the transformation of their plant was reached, they were ready to focus more aggressively on sales.

Among their new product lines is 'tumbled old Chicago' [measuring 2 x 3–1/2 x 8 inches], recycled brick categorized as *seconds* because they have broken corners. After these brick are put through a tumbling process in a revolving cylindrical drum called a *tómbola* [*tumbler*] that rounds off broken corners, they are essentially transformed into *firsts* and are sold to satisfy demand in a market niche for antique Chicago–style brick. In one sense the tumbled old Chicago brick in the Reynoso line grew out of more exacting sorting practices instituted at the new plant. Previously, brick that come from the lower reaches of the kiln where the fire is hottest came out of firing with weakened or broken corners. This was recognized by the Reynosos as being in the 'seconds' category but was, nevertheless, bundled and sold together with unflawed brick without informing clients of the practice. After the new plant was up and running, the Reynosos decided to end this deceptive practice. Now damaged brick is manually separated out prior to bundling, either to be sold as seconds, used in kiln repairs, or otherwise recycled.

In the Texas and the greater U.S. brick markets, the term 'Chicago brick' refers to used brick mostly reclaimed or salvaged from historic building demolition projects in the Midwest and Northeast. Many of these brick are supplied by the *Colonial Brick Company* located in Chicago that advertises itself as "America's Largest & Oldest Antique Chicago Brick Supplier." One of this company's principal products is "Antique Chicago Common Brick" that comes in pink or buff colors and comes from salvage projects in the city of Chicago. Nevertheless, used or reclaimed brick sold under the Chicago category may come from salvage projects elsewhere in the Midwest, Northeast or, for that matter, the Southeast. About ten percent of the Reynoso's business involves this special category of brick.

With the intent of transforming their inventory of seconds into a simulated Chicago brick, the Reynosos started experimenting with methods through a trial and error process to round off the jagged or partially broken corners of their inventory of seconds through the above–described tumbling process. Their initial effort, however, involved loading seconds onto a conveyor belt tilted upwards, so that at the end of their ride up the conveyor the brick would tumble off onto the ground below. This did not work because the brick needed to pass through the process ten or fifteen times to obtain the desired effect. Also, falling brick would damage brick already on the ground, creating more waste. After abandoning the conveyor belt method, the Reynosos tried loading brick into a dump truck with a backhoe and then repeatedly dumping them. This did not work any better than the conveyor belt method. In short, the experiment lasted several months before resulting in the improvisation of a motor–driven, revolving cylindrical steel drum with a loading door mounted on a steel frame. The revolving action causes brick to tumble against each other inside the drum and creates in a reasonably uniform way the desired rounded–corners appearance.

Another factor that motivated Tony to pursue the tumbler project was his own aesthetic preference for the antique look in building materials. His personal aesthetic inclined him toward salvaged or reclaimed brick, and raised for him the question of how their plant could make a brick that appeared older than it really was. When their brick come from the kiln they seem to him very clean or very new in appearance and he wanted to find a way "to make them appear like the brick used to build the kilns. That is, you look at them and they appear to be beaten up, they look wasted—the wind and the rain have eroded them little by little." Tony was determined to find a way to mass–produce this effect.

The Reynoso's success with their tumbler project has enabled them to ingeniously transform seconds from their own production into a product that simulates the appearance of authentic antique Chicago brick. Their success in this market niche, however, reflects not only the appearance of their product but the very competitive price at which they can sell it. Tony explained that reclaimed or salvaged old Chicago brick shipped from the Midwest, arrive in the Rio Grande Valley with a price of five hundred fifty dollars per thousand. Valley brick dealers, themselves included, have to sell Midwest Chicago brick for $625 per thousand to make a profit on it. That makes it almost twice as expensive as their own regular new king size brick which sells for $325 per thousand and is somewhat larger than the Chicago [3 x 3 x 9 inches vs. 3 x 3 x 8]. By contrast, the Reynosos are able to sell their version of old Chicago for $450 per thousand ($0.45 each). Since much of the Midwest Chicago salvage brick has to be hit with a mason's hammer to remove old mortar, their appearance ends up being quite similar to that of the Reynosos' tumbled brick but at a much higher price. Another market advantage of their tumbled brick is that, according to customer color preference, they can be dipped into a paint solution prior to being tumbled so as to produce a mottled or blotched antique look, with the chosen color dominant.

One trend in the binational building market over the past decade or so has been the decline of the use of the traditional *adobón* by masons in Reynosa who have turned mainly to cement block construction coupled with its increased use in the Rio Grande Valley. Fifty percent of Reynoso's sales in the Valley are *adobones* modified in size to conform to the Southern Building Code that requires a minimum thickness of 2–5/8 inches in wall brick. The traditional *adobón* measured 2–1/4 x 10–1/2 x 5–1/4 inches, whereas the new Reynoso version measures 3 x 10 x 5 inches. The reason for the increase in use of this brick in Valley construction has to do mainly with cost savings accruing to the builder. This brick is mortared onto walls sideways [*de canto*] with the sand struck or rough side exposed, rather than lengthwise with the smooth size exposed. Although *adobones* are more expensive [$0.15 each dealer cost/$0.60 retail delivered in Valley] than standard size 3 x 3 x 9 inch [$0.08 each dealer cost/$0.32 retail delivered in Valley], the difference is more than offset by a 25 percent cheaper cost per square foot of building with them due to their larger size. Moreover, less sand and mortar are used in building with these larger brick, hence, providing another cost–saving incentive to the builder.[3]

The *adobón* has become so popular among builders in the Rio Grande Valley that in the fall of 2007, it represented 50 percent of Reynoso plant sales. It is not surprising then that they have diversified their output to include a large size [3 x 8 x 16 inches] compared to the regular size [3 x 5 x 10]. Both sizes are produced in pink, cream, and yellow colors, both natural and whitewashed. Only regular size *adobones* are available with the tumbled look.

Aside from two categories of *adobones*, the Reynosos produce three other categories of modular or common brick: eight–inch [3 or 4 x 2–3/8 x 8], nine–inch [3 x 3 x 9], and bars [*barras* measuring 3–1/2 x 2–1/4 x 10–1/2 inches]. In addition, they produce pavers in four different sizes, and four types of specialty brick. Considering combinations of possible colors, styles, types and sizes the new Reynoso plant's product line includes 220 different brick.

Since all of these brick are produced for the Texas market, the Reynosos' investment in a new plant with increased capacity required them to establish a yard and sales office/showroom on the Texas side of the border where an inventory of packaged brick can be stored for immediate shipment and customers can see the product line on display. This was a big step in the transformation of the Reynoso's business since the old plant sold brick only to clients who either showed up at the plant or ordered by phone and arranged for their own transport from the plant. This was in accordance with Toño Reynoso's philosophy, modeled after that of his elder brother Sebastian, that their problem was to produce brick and that, once produced, the brick would sell themselves. In other words, from his perspective, it was not necessary to spend on marketing since plant inventories were, sooner or later, purchased by clients with whom business had been regularly conducted year after year.

A large, fenced site was leased at a strategic and easily accessible location along a main road in Pharr close to the new international bridge as well as to the Military Highway that connects directly to the old bridge in Hidalgo. A large, attractive bi–directional sign, mounted on two tall metal posts along the road, displays in large letters, *Ladrillera Reynoso, The Handmade Brick Company, Desde [Since] 1950* with the company's logo in the center and its phone number at the bottom. The site has a modular building that serves as a combination office/showroom and adequate open space for brick storage, loading, and vehicular parking. The Reynosos own two flatbed tractor–trailers, one of which hauls brick from the Reynosa plant to the Pharr yard, and another used to deliver brick to customers in the Valley. Brick are unloaded and reloaded at the yard with a forklift. Aside from Tony, who now spends most of his time focusing on sales, the Pharr operation has two full–time employees, one in sales and one who drives the trucks and operates the forklift. The monthly cost of this operation is about $15,000 [See Appendix 1. Photos 90–95 illustrate the tumbler area at the

Reynoso plant, including the machine and its products, as well as the tinting process. Photos 96–98 show the yard and sales office in Pharr].

Plant Output and Wages: Time Series Analysis, 2005–2007

Data obtained directly from a sample of Reynoso plant records for January, June, and October—a total of thirty eight weeks—over a three–year period (2005, 2006, 2007) provide a cross–section through time of output and worker salaries. These three months were selected to cover the extremes in Reynosa's weather, especially in terms of temperature: from relatively cold/cloudy conditions in winter [average temperature sixty–one degrees F], to sunny/hot conditions in summer [average temperature eighty–six degrees F], and to generally temperate conditions during the fall. Heavy rain is possible in any of these months but is most probable during January and least probable during October. The new Reynoso plant has roofing over clay storage, as well as over brick molding and drying areas to make it less susceptible to slowdowns due to heavy rainfall, cold and rainy weather. Nevertheless, bad weather still has a negative impact on general working conditions at the plant and inevitably reduces output.

Plant records organize production data in ten columns with a line or row for each employee by name. The four columns on the left side of the tabulation sheet are 1) number of days worked; 2) day's wage; 3) subtotal; 4) loans during the week. The right side of the sheet is devoted to transactions at the company store and includes purchases of soft drinks and snacks. These data present a comprehensive picture of individual worker productivity and income, as well as output and expenditures for the plant. As Table 9.1 shows, over the thirty–eight week period the plant's brick output (all types) was 937,528 pieces for a weekly average of 24,671 pieces. Not shown in this table is an interannual variation in average weekly output from a high of 31,314 pieces in 2005, to a low of 16,861 pieces in 2006, and to 26,351 pieces in 2007.

The range of weekly output over three years was from lows of 2,710 and 5,660 pieces in January, 2007 to highs of 45,830 pieces in October 2007 and 44,770 pieces in June 2005. These weekly output fluctuations are reflected in monthly output averages shown in Table 9.2 The January 2007 low of 7,665 reflected a combination of absenteeism among key molders and their participation in kiln firing. More striking than these highs and lows in the output series, however, is the absence of any week without production. Despite changes in weather conditions, turnover in personnel, equipment breakdowns, or orders, there was always weekly production from a minimum of three to a maximum of ten molders.

The relatively low total output of just under 220,000 for the thirteen weeks sampled in 2006, shown in Table 9.1, together with the low monthly average output figures shown in Table 9.2, probably reflect worker turnover, especially among molders, more than any other single factor. Of the plant's total of twenty–two employees on the payroll in January, 2005 only five were still on the payroll of eighteen employees in October, 2007. Three of these were molders. More than half of the employee turnover occurred during 2006, when four new molders were hired and trained. The average number of molders employed weekly declined from 8.5 in 2005 to 5.4 in 2006 but rose to 10.5 in 2007. This highlights the importance of molders to the plant's output performance. Weekly total productivity per molder was also lower in 2006 [3,131 pieces] than in 2005 [3,998 pieces] and 2007 [3,764 pieces].

The importance of molders to plant output is only partially reflected in the payroll data shown in Table 9.3. The molders' (M) average weekly wage over the three–year period is only slightly higher than that of general workers, but is less than that of kiln stokers—underscoring the critical importance and specialized nature of that job to the production process. During 2007 stokers were paid anywhere from 250 to 385 pesos per turn, with the highest salary paid to the most experienced stoker. All employees in the general category (G) were paid a wage for a 6–hour day ranging from 170 to 316 pesos. The highest paid employee was Luís Escobedo, the plant foreman; the clay mill operator is the next highest paid at 275 pesos, and the office manager earns 250 pesos daily.

Wages were paid hourly for longer or shorter days. The minimum hourly wage was 25.50 pesos [$2.55] in 2005 but rose to 28.30 pesos [$2.83] in 2007. This increase paralleled an increase in the average weekly wage per worker as Table 9.3 shows. The anomaly in the series, represented by the 2007 wage for molders, can be explained by the attrition factor explained above that clearly resulted in higher productivity from the molders with longevity.

Among the workers with whom I discussed the topic, I found only one molder who expressed discontent regarding the payment situation at the plant. He was a relatively recent hire with long prior experience in a Camargo plant that, for many years, had a unionized workforce. He considered it to be advantageous to be employed under a union contract that was periodically renegotiated to include wage increases. Others, however, emphasized that they were satisfied with their salaries at the Reynoso plant, and had found it to be sufficient over the years to adequately feed, clothe, educate, and otherwise provide for their children.

As the data presented in Table 9.4 for the months of January, June and October in 2005, 2006 and 2007 demonstrate, a decline in output from 2005 to 2007 was accompanied by a 25 percent increase in the wages bill that translated into a $.03 cost increase per brick. Still, more detailed analysis indicates that the substantial drop in output from 2005 to 2006 shown in Tables 9.1 and 9.2 corresponded to a drop in the average number of molders employed weekly in those two years from 7.9 to 4.6; and that this trend was reversed in 2007, even though

Table 9.1
Number of workers, days worked and total output at *Ladrillera Reynoso* over 38 Weeks in 2005, 2006, and 2007

Year	Number of Workers By Category			Days Worked		Output
	G	K	M	K	M	
2005(12 wks)	193	19	94	925	79	375,770
2006(13 wks)	167	23	70	826.5	48	219,194
2007(13 wks)	193	44	91	1024.5	109	342,564
TOTAL	491	75	227	1954.9	217	937.528
AVG/WEEK	14.4	2.2	6.7	57.5	6.4	24,671

Table 9.2.
Monthly average output for January, June, and October in 2005, 2006 and 2007 at *Ladrillera Reynoso*

	2000	2006	2007
January	32,308	24,800	7,665
June	32,542	14,415	32,044
October	29,092	11,980	37,921
TOTAL	93,942	51,195	77,630
MEAN	31,314	17,065	25,877

Table 9.3.
Average Weekly Wage Per Worker (U.S. Dollars) By Category for 38 Weeks in 2005, 2006, 2007 at *Ladrillera Reynoso*

Year	Worker Categories General (G)	Kiln (K)	Molding (M)
2005	943.70 ($94.37)	1031.94 ($103.19)	911.52 ($91.15)
2006	1049.43 ($104.94)	1237.78 ($123.78)	1284.14($128.41)
2007	1069.21 ($106.92)	1256.29 ($125.63)	1078.41($107.84)
Totals	3062.34 ($306.23)	3526.01 ($352.60)	3274.07($327.41)
Mean	1020.78 ($102.08)	1175.33 ($117.53)	1091.35($109.14)

Table 9.4.
Wages Bill, Output, and Cost Per Brick: 3 months in 2005, 2006 and 2007 at *Ladrillera Reynoso ($U.S.)*

Year	Wages Bill	Output	Wages Bill/Output
2005	$28,742	375,770	.076
2006	$29,362	219,194	.134
2007	$35,977	342,564	.105

total output was still less than in 2005. The increase in output in 2007 corresponded with an increase in the average number of molders employed to 7.0 compared to 5.4 in 2006.

Further computations, not presented in the tables, show that the average wage per worker increased from 939.30 pesos in 2005 to 1096.71 pesos in 2007, while the number of workers on the payroll also increased from 306 to 328 over the same period. The average wage per worker, however, was higher in 2006 than in 2005 and 2007. This seems to reflect the much higher average weekly output by molders in that year of 5,218 compared to 3,998 in 2005 and 3,764 in 2007. Higher than average output translates directly into higher wages.

Of course, the plant incurs many other costs aside from labor, including housing, utilities and social security insurance, that increase by a few cents the average cost of production per brick. Also, the cost of transporting and marketing brick also must be taken into account. This latter cost is estimated to average $15,000 per month in 2007 or about $0.13 added to the per unit cost. Adding another two cents for additional labor costs brings the estimated per unit total to $0.255. Even at this elevated per unit cost, sufficient cushion is provided by the average selling price per brick in the Texas market to provide for a profit margin. For example, at October 2007 market prices of $325 per thousand for standard eight–inch [2 x 3–1/2 x 8"] or $450 for Tumbled Old Chicago, the $0.32 and $0.45 per unit prices provide substantial cushion to absorb costs and earn profits in view of the difference between the relatively low overall per unit labor cost of $0.255 and the per unit selling price.

Finally, the plant's payroll data indicate that loans were made during thirty–three of the total of thirty–eight weeks between 2005 and 2007 to an average of five workers weekly. The average amount loaned was 122 pesos [$12.20], but the range per loan was from 20 pesos [$2] to 2900 pesos [$290]. This loan service was most used among the category of hourly workers, and was relatively little used by the more highly paid stokers and molders. However, molders had by far the highest average amount per loan at 313.33 pesos or $31.33. This compared with 113.50 pesos [$11.35] for general hourly workers and 128.33 pesos [$12.83] for stokers]. This loan policy reflects a long–standing tradition of employers in the border brick industry and is in accord with the culture of patron–client relations in Mexican commercial–industrial culture (Cook 1998:111–123).

Luís Escobedo: the Youngest *Mayordomo*

Only thirty–one years old when I interviewed him in 2008, Luís Escobedo is unusually young to hold the job as foreman [*mayordomo*] in a border brick plant, a job typically assigned to older, more experienced workers. When the Reynosos offered the job to him in 2006, he had worked for them for only one year but they quickly recognized that he had an exceptional aptitude for many of the key jobs at a brick plant like clay preparation, molding and setting. He was born into a family of Los Longoria brickmakers, and his father and two elder brothers worked for many years in Juan Tueme's plants. Luís, however, did not learn brickmaking from them and worked in the local cardboard recycling business prior to taking his first job in the brick industry with José María Cantú at the age of seventeen. He began loading trailers, then turned to setting unfired brick in the kiln, and, finally, worked as a clay preparer and molder. His natural aptitude was most apparent here since he was making excellent brick after only two weeks, whereas most apprentices require at least a month to master these tasks—that is, if they do not abandon the job first. Luís worked for eleven years at the Cantú plant prior to going across the river in undocumented fashion to work in the Dallas construction industry for a year as a mason's assistant where he noted ironically, "I was laying factory–made brick" [See Appendix 1. Photo 83].

His first job at the Reynoso plant was to make *adobones* but he also worked as a kiln setter before being offered the job as foreman. This is the highest paid job on the plant's payroll, averaging one hundred sixty dollars weekly in 2006 and two hundred dollars in 2007 with housing and utilities provided, but requires a six–day week and multiple responsibilities. Luís envisions it as a lifetime job; he likes the work and considers the plant to be a good place to raise his three children. His expectation is that his nine year old son will also work at the plant after completing his education but he noted that will be his son's decision.

Routine responsibilities of plant foreman include supervising clay extraction and processing, training new employees, supervising kiln setting and drawing, and general troubleshooting. He does not have a supervisory role in kiln firing because this is the province of the Reynosos themselves and of their highly skilled stoker/molder, Pedro Escamilla. In any case, on any given day Luís can be seen traveling on a bicycle between different areas of the plant checking on work in progress. At the time I interviewed him he was supervising the preparation of a storage area near the kilns for the arrival of a load of recycled pallets used as fuel. His next visit that afternoon would be to the tumbler area to supervise the transport of materials to be colored and tumbled.

Luis's troubleshooting role came into play in late 2007 when the employee in charge of the clay mill failed to show up to work one morning, essentially quitting without notice, resulting in a production crisis at the plant. Luís played a key role in performing some of this employee's duties himself, shuffling existing personnel to temporarily cover these job duties, and in training a new employee to permanently fill the job. This 2007 mill crisis exposed one of the drawbacks to mechanization of clay preparation, namely, to bring the entire production process to a halt due to a malfunction or other problem in clay processing. In unmechanized plants where every molder prepares his own clay, the absence of one molder will reduce but will not stop production as it does in a mechanized plant.

Apart from supervising the training of new molders, the category where jobs are most likely to be available for new hires, Rafael admitted that little supervision is required of molding. The main concern is to check daily that clay being delivered from the mill to the molding area meets requirements in terms of composition, for example, the proportions of clay and sand; consistency, for example, the absence of gravel or other foreign matter; and plasticity, that is, water content. This monitoring process begins at the clay pits where he must check regularly to make certain that the material transported to the mill has the right proportions of clay and sand content.

New employees at the Reynoso plant are usually hired in response to ads placed in the local newspaper. If they have no experience in the brick industry, they are first put to work at the easiest job which, according to Luís, is loading unfired brick into wooden containers for transport to the kilns [*"llenar cajones con ladrillo crudo es lo más fácil"*]. Before that job or most others at the plant can be done properly, however, Luís must teach every neophyte "how to grab brick" [*como agarrar ladrillo*]; until they learn how to handle brick properly, according to him, "they do not advance" [*Si no se enseñan agarrar ladrillo pues no avanzan*]. Once a new employee spends several days learning these simple tasks, he is then apprenticed with an experienced worker in jobs like setting, drawing, or coloring.

Training of new molders is among the most challenging tasks Luís is required to perform at the plant. He admits to only limited success in this activity: "It is very complicated to teach them to do this in a reasonable time. We have tried to do it but they find it to be very hard. And they do not last long at the job. They go away. You have to start them out with less tiring jobs. The job of brickmaking is very tiring." Despite these difficulties, the fact that the Reynoso plant has a young, dedicated foreman like Luís Escobedo, together with the demographic growth of Reynosa that provides a pool of young, if inexperienced, potential workers, suggests that the plant will have the capacity to replenish or expand its labor force in the future.

Negotiating a Dual Currency Economy

As discussed in chapter 1, every enterprise in the border economy that produces commodities in Mexico for export to the United States must develop an accounting strategy to address the realities of conducting business in two currencies, the U.S. dollar and the Mexican peso. The Reynoso family enterprise today follows business strategies and practices derived from decades of experience with transactions in the border region's volatile dual currency system. The first principle of this conduct is protection against peso devaluation by keeping monetary assets in dollar accounts in banks on the U.S. side of the border. The other side of this principle is to keep only a limited fund of short–term operating capital in peso accounts in Mexico. Although it is now possible to have dollar accounts in Mexican banks, there is a disincentive to do so for border businesses, given the easy accessibility of banks in U.S. territory, and in view of the possibility of Mexican government intervention that might have negative impacts on the value and freedom of use of such accounts.

The Reynosos necessarily maintain a peso account in Mexico to pay weekly operating expenses such as salaries and fuel costs. The account is replenished at the start of each month but never exceeds anticipated operating expenses for that month. This amount may vary when, for example, taxes must be paid in Mexico for a durable capital good such as a vehicle or machinery purchased in the United States for use in Mexico.

Each truckload of brick that is transported from the plant in Reynosa to the yard in Pharr is invoiced in pesos for purposes of satisfying requirements of the Mexican government and the customs process but clients in the U.S. are billed in U.S. dollars. These dollar payments are then exchanged into pesos and temporarily deposited in a peso account in Mexico. This has to be done by Mexican export producers to demonstrate to the Mexican government that there is a correspondence between the invoiced price of exported commodities and the amount received in payment for those commodities. In the Reynosos' case, after taxes and expenses are paid, the surplus remaining in the peso account earmarked as profits is converted into dollars and deposited in a dollar account in a Texas bank. Tony summed up the rationale for this series of currency exchanges as follows: "The surplus does not stay in pesos because we run the risk of a peso devaluation and of having less at the end that we had at the beginning." [*No se queda en pesos porque correámos el riesgo de una devaluación de los pesos y quedamos con menos al final*].

Although the Reynosos are obliged to conduct business in two currencies, their accounting and planning activities revolve around dollars. As Tony explained:

> In reality, for us, even though we participate in peso transactions, we always think in terms of dollars [*Nosotros realmente, si bien manejamos el peso, siempre pensamos en dólares*]. How much does a thousand brick cost us? To manufacture a thousand brick costs us so much in dollars. How much does a worker earn? How much does fuel cost? We always think in dollars, so if we have to lose money in the conversion of dollars to pesos and then afterwards of pesos into dollars, that is preferable to the risk of a devaluation.

One Borderlander's Views on Identity in A Binational Region

Américo Paredes argued forcefully that, regardless of their birthplace on the Mexico or Texas side of the border, the Spanish–speaking people of the lower Rio Grande/Rio Bravo region "thought of themselves as *mexicano*" (1993:26). I have explored this thesis among border brickmakers and, with minor qualifications or exceptions, found it to be true (Cook 1998:46–54; 2001:35–36). This exploration is widened and deepened, especially with regard to the complex relationships between nativity, upbringing, travel, work experience, bilingualism and identity through the case of Tony Reynoso.

Tony's father, Toño, whose permanent residence was in Reynosa's *Colonia Aquíles Serdán*, arranged to have all of his children born in the same private clinic in Pharr to assure them of being U.S. citizens. He also arranged for Tony and his siblings to regularly visit relatives in the Rio Grande Valley and encouraged them to gain work and other experience there. Nevertheless, they lived and were educated in Reynosa and, as noted earlier, Tony received his degree in architecture in Monterrey, Nuevo León.

Tony admits that he spoke very little English until he went to Dallas after finishing his degree. After spending several years there he has achieved bilingual status. He considers this as necessary for the conduct of his business and for living successfully as a borderlander. In his words: "The border is another world and only a bilingual person can know it well. Being bilingual opens up another world. We are in two territories that we know very well. There is no marked dividing line between English and Spanish [or between *Tejanos* and Mexicans]. We are them and they are us."

Tony's identity as a bilingual borderlander is under girded by his strong sense of being Mexican. He expressed it this way: "Look, personally I consider myself Mexican. And I will explain why. My being [*ser*] is Mexican. I eat Mexican. I dress Mexican. I look Mexican. I do things that all Mexicans do, even if they were born in Japan." His thoughtful approach to the identify question has not only benefited from his lived experience on both sides of the lower border but also through his work experience in Dallas, his extensive travel in Mexico and the U.S., and his visits to Argentina, where his wife was born and raised.

Tony resides with his wife and children in Mission, Texas and for the past several years has commuted daily to his plant in Reynosa, usually over the bridge at Hidalgo, Texas. Even during this post–9/11 period with increased border security, he has never been asked to produce documents, much less stopped and searched, by U.S. customs. In fact, as my trips across the bridge with him attest, he is well–known to the crossing guards and is often greeted by name and asked how his business is going. As he expressed it: "I cross the border into the U.S. without ever having the need to produce any document to prove my citizenship."

So, born in the U.S.A. and, therefore a U.S. citizen, raised and educated in Mexico with periodic visits to the Rio Grande Valley while growing up, self–identified as a Mexican, yet carrying a U.S. passport, and admitting that he is a "person who is very grateful to . . . . [the U.S.]," Tony Reynoso is very much a borderlander. He is aware that "the situation in which we find ourselves on the border has served me very well to enjoy a lot of creature comforts [*comodidades*], and that "we enjoy a geographical position that gives us the opportunity of managing two different cultural surroundings [*entornos*] or environments [*ambientes*]." He elaborated: "We know the traditions of Mexico, the way to behave over here, and the way in which we have to behave over there [U.S.]. I have never felt uncomfortable on the U.S. side of the border, not even in Dallas. I felt very comfortable even there."

When I asked him how he compared life in the Valley with life in Mexico, and about his attitude regarding '*Tejano*' identity, he responded as follows:

> The Valley does not belong totally to Mexico even though the people are definitely Mexican. I am very proud to be from the region, a region that has a border. As things stand, as a child, although I did not know English, I came a lot on weekends and on vacations to the flea markets and to visit relatives in the Valley. Over the Christmas holidays and during the summer I came over to work with my cousins in carpentry all over the Valley. So, I also grew up in the Valley and I know it well.

Even so, Tony hesitates to call himself a *Tejano* and prefers instead to think of himself as "of the region because it has two sides." The bottom line for him, however, is that his basic make–up [*forma de ser*] is more compatible with northern Mexican culture than with the culture of *Tejanos* born and raised in Texas. From his perspective, his experiences growing up in the Valley were not sufficient to give him *Tejano* identity. Aside from differences in food, he also views *Tejanos* as different culturally in the sense that they perceive *respect* [*respeto*] in a different way than Mexicans do. He sees them as being generally less respectful than Mexicans in their social relationships.

His self–concept as a "person of the region" involves the notion of a borderless border: "I consider myself a person of the region without putting a border in it" [*Me considero una persona de la región sin poner una frontera*]. He tried to drive home his point by asserting that he is as equally content at the brick plant or in the main square of downtown Reynosa as he is at his place of business or at a shopping mall in Pharr. In short, he "feels at home" [*Me siento en casa*] on either side of the border, and contrasts this with not feeling at home when he visits his wife's native country of Argentina.

When I asked Tony in 2007 what his expectations were regarding the announced tightening–up of border security by the U.S. Department of Homeland Security, he admitted that his citizenship documentation might become more important at border checkpoints. Given the imminent anticipated change and the controversy surrounding the immigration issue in the U.S. and Mexico, he was weighing the option of seeking dual citizenship since the Mexican government grants Mexican citizenship to any individual, like himself, born outside of Mexico to Mexican parents. He hastened to add, however, that as of the date of that interview in 2007, he had never needed to prove his Mexican identity upon entering Mexico at border crossings. That all changed, however, perhaps forever, when both Tony and I, on our return to Pharr after visiting his plant in Reynosa on January 30, 2008, were required to produce documentation of citizenship at the U.S. Immigration checkpoint when we crossed the Reynosa–Hidalgo bridge in his pick–up.

The implication of Tony's mention of this dual citizenship option, especially in view of the reciprocity built into cross–border immigration policies, coupled with the historical fact that such changes are typically initiated by the U.S. government and reciprocated by the Mexican government, is that the prudent strategy for a binational commuter and borderlander like himself is documented dual citizenship.[4]

Trends in Texas Brick Manufacture and Implications for the Future of Mexican Handmade

For all of the ceramic engineering innovativeness of large–scale, high–tech brick manufacturers and brickmaking machinery manufacturers, the simulation of genuine artisanal brick presents formidable challenges. Admittedly some of these manufacturers, through the development of innovative mechanized processes incorporating wood and/or steel molds and antiquing methods, have made great strides in producing brick that have many of the markings and linear irregularities of artisanal brick—even though human hands do not touch them during the molding process and they are not demolded onto a flat dirt surface as are handmade brick. The key to this simulation process lies in the proper mix of higher sand–content and lower sand–content clays; molds are sanded before clay is put into them as are the exposed surfaces of the clay already in the molds. Used in these ways, a controlled mix of dry and wet sand imparts to each mechanically molded brick the rough texture, markings, and irregularities of shape and texture associated with hand molded brick, and also facilitates demolding [ejection of the wet, just–molded brick from the mold]. Brick made in this fashion are referred to variously as machine–made sand molded, wood mold look–alike, soft–mud molded, old colonial face brick, and so on.

As discussed in chapter 2, beginning in 1913 with a small artisanal brick plant on a ranch named *El Gavilán* on the banks of the Rio Grande south of Mission, Texas, a German immigrant, Guenther Weiske, subsequently relocated his plant to a 60–acre site in Madero closer to Mission. This became the main plant of the Valley Brick and Tile Company that later acquired another plant on a 160–acre site in Rio Grande City [Starr county]. Valley Brick produced tens of millions of hand–molded and machine–made [soft–mud, pressed and stiff–mud] brick under the banner of "What the Valley Makes, Makes the Valley." Today, these plant sites, abandoned in 1972 and 1980 respectively, stand as vivid reminders of deindustrialization, and as a metaphors for what was, for more than a century, the principal non–agricultural industry in the lower border region.

There are complex reasons for the shutdown of the Madero and Rio Grande City plants which, as indicated above, were both owned and operated by Guenther Weiske from 1943 until 1965 when Rudy Nordmeyer assumed ownership of the Rio Grande City plant, Rio Clay Products. Both plants had extensive on–site clay deposits, commitments to technological innovation, and a local labor supply. Probably, their demise lies in a combination of factors including bad luck and timing, relatively long distance of plants from key urban markets and, most importantly, economies of scale and the trend toward consolidation [and, ultimately, globalization] in the Texas brick industry. Just at the time that these wider market forces were causing attrition and consolidation among Texas brick manufacturers, Valley Brick bucked the trend by de–consolidating, hence losing whatever advantages it might have had through economies of scale. This made Weiske's Madero plant and Nordmeyer's Rio Grande City plant more vulnerable to the rigors of the capricious and competitive Texas brick market.

The Austin–based Elgin Butler Brick Company had a long–standing business relationship with Valley Brick and Tile for the distribution of Butler brick in the Valley and the supply of handmade brick, much of it from Reynosa, to the central Texas market. The Butler family's company was itself ultimately transformed by the same market forces that impacted the Valley Brick and Tile as indicated by the 1991 sale of its Elgin plant to Acme Brick. The Elgin plant itself originated in the first decade of the twentieth century as Prewitt Brick Company operating with two hand presses and wood– and coal–firing scove kilns; it was acquired by the Butler family in 1965, after having experienced several earlier transformations in scale and technology under the name of Elgin Standard Brick Manufacturing Company. Today, in keeping with these historic trends toward mechanization and consolidation that have intensified in recent decades owing to globalization, the Texas clay brick market is dominated by three international corporate behemoths, Boral Brick, Hanson Brick, and now Acme Brick, that in 2000 was acquired by Warren Buffet's Berkshire Hathaway conglomerate.

Until recently, Texans did not have to look far for an exemplary, if little publicized, case of a large–scale manufacturer engaged in the production of simulated handmade brick. The automated, robot–equipped Acme soft–mud plant in Butler, Texas [just outside of Elgin and built by the German firm Lingl] during the 1990s and until 2006 produced an average of 46 million molded brick annually [typically 8–3/4–inch length x 3–inch width x 3–inch depth] in a variety of colors ranging from red to pink to near–buff. Their production involved a soft–mud process incorporating a JC Steele mixer, two DeBoer mixers, and De Boer mould chain equipment and automatic brick press with eight Hubert filler heads. The soft–mud was produced by mixing ground and sifted clays mined at the plant site into a blend at a 60/40 percent ratio of silt and sand to clay. This blended, water–saturated clay then passed by conveyers into the molding system, with a capacity of 24,000 pieces per hour. This system incorporated a series of brick molds that were sanded prior to filling.

Next, the sanded and clay–filled molds were demolded onto drying boards and transfer–loaded into rack style dryer cars that were conveyed by an rail system into a series of eleven gas–fired drying chambers [33.792 brick per chamber; 50 percent of the heat recycled from the kiln]. Once dried from 26 percent to 1 percent or less moisture content, the brick were unloaded and robotically reset into kiln cars on rails that moved into a line–up for entry into the gas–fired tunnel kiln that operates continuously. Each car held 11,340 brick and twenty–five cars filled the kiln's capacity of 283,550 brick. Within the kiln that is 386 feet long loaded cars moved forward in indexed surges or pushes, rather than continuously, through a fifty–hour firing cycle at temperatures reaching a maximum of 2140 degrees F. The brick emerged from the kiln cooled down and ready for robotized unloading from the kiln cars, stacking, and strapping. The tunnel kiln had a daily output of 135,000 brick.[5]

The entire computer–controlled process from dry, unmixed raw material to fired brick—stacked, strapped and ready for fork lifting to the inventory yard—required about four and one–half days for a daily kiln output of 125,000 brick. This process was by no means trouble–free, however, and was relatively high cost in terms of gas, electricity and water consumption. Short stoppages were common during the brick molding, dry setting, and robotic packaging stages. Workers had to intervene to clear away excessive mud droppings or clogging, to replace broken cutting wires in the vertical extrusion system, or to pick–up, re–arrange, or toss out damaged brick during the setting and packaging stoppages.

The variety of molded brick produced by this process at Acme's Elgin molded brick plant included the best–selling "Charlestown", "Castle Rock" and "Williamsport" lines. These lines proved to be quite popular in the custom–building markets in Houston, Austin, San Antonio and surrounding areas on jobs that are not bound by requirements to replicate, rather than simulate, an authentic, traditional, handmade look.

This is a important qualification because, as any architect working in the lower Rio Grande Valley will attest, the only possible source of replacement brick for historic reconstruction purposes, or for custom projects stipulating hand molded brick, are the extant artisanal brick plants on the Mexican side of the border that mold brick the old fashioned way. That is, with workers who knead clay as if it were dough, put globs of it into sanded wooden molds, disperse it into the mold and level it off with a wooden leveler or by hand, empty the mold onto a patio or drying surface that is smoothed–out ground. The natural dirt surface of the drying floor imparts to hand–molded brick, that is typically canted, skintled or turned on edge to facilitate drying, its own rough texture and imperfections. Moreover, the clay used in these artisanal plants is from the same riverine sources used in past centuries, and the firing technology, employing mesquite or other kinds of wood, is also relatively unchanged. Other things remaining equal, even the gas–firing of unfired brick made by hand from this same clay, as opposed to their wood–firing, may not yield the precise color or look required by a particular restoration or custom project. [See Appendix 1. Photos 99–112 show facilities, activities and scenes at the Acme plant in Butler, Texas corresponding to the descriptions given in the text].

Conclusions

It is reasonable to suppose that O'Neil Ford would have been appreciative of Acme's experiment with machine–molded soft–mud brick at its Elgin plant, and would have been enthusiastic about the persisting and innovative hand–molded brick industry in Reynosa. Ironically, given the competitive pressures operating in the Texas brick market, combined with the lingering inefficiencies and relatively high cost per unit of automatically molded soft–mud brick, that inevitably have negative cost accounting implications, in the summer of 2006 Acme eliminated its soft–mud operation and converted it completely to more cost–efficient and less problematic non–molded, stiff–mud extruder technology.

This means that the future production of rough textured, capriciously marked, irregularly shaped, adobe–looking clay brick for the Texas market now lies exclusively with small, non–automated but partially mechanized, artisanal plants like those in Reynosa. This market niche caters to custom builders and their architects and, generally speaking, is more demanding in terms of technology and quality control than was the mass, speculative building market that was the engine of growth for the Mexican handmade industry during the second half of the twentieth century. The clay brick segment of the speculative building market today seems to lie predominantly in the hands of the Big Three manufacturers—Boral, Hanson, and Acme.

So long as demand for artisanal brick exists, however, and given the locational and transport economics of clay brick production, it is likely that handmade brick manufacture for the Texas market on the Mexican side of the border will continue. Their main competition is from suppliers of Old Chicago salvage brick and from a handful of plants in the southeast U.S. that still employ soft–mud technology to produce molded brick that simulate authentic handmade brick. The case study material from *Ladrillera Reynoso* suggests that it can successfully compete in the Texas market, so long as new molders can be trained and retained. Given the combination of motivation, special aptitude and physical fitness required of brick molders, and the rising expectations of the Mexican labor force regarding pay, benefits, and working conditions, it is not certain that the availability of brick molding labor in Mexico will last as long as consumer demand for handmade brick in Texas.[6]

Notes

1. Gurcke (1987:102–103) discusses the issue of distinguishable strike marks on soft–mud brick but observes that the movement of a wooden or metal striker across the top of a filled mold to remove excess clay may not leave distinguishable marks on handmade brick. In the case of border Mexican handmade brick, even if such marks might be detected immediately after the striking occurs, they are obliterated by marks produced when the mold is emptied onto the ground to begin drying.

2. These kilns are capable of reaching temperatures as high as 1500–1600 degrees F which is still below the maximum temperatures reached in state–of–the–art kilns fired with natural gas, as for example at the Acme plant in Elgin, Texas where temperatures as high as 2140 degrees F are reached. Vitrification or the stage when the brick mix changes chemically into a ceramic material usually occurs at temperature ranges between 1400–2200 degrees F but may occur at lower temperatures with certain clays and clay mixes.

3. The reader is referred to the excellent discussion in Campbell and Pryce (2003:14–20) of the necessary and intimate linkages between brickmaking and bricklaying.

4. I have borrowed the term 'borderlander' from Oscar J. Martínez (1994:Ch. 1).

5. A concise, instructive account of the industrial technology of brickmaking can be found in chapter 2 of Gurcke's (1987) handbook. See Searle (1956) for the most detailed, comprehensive and authoritative treatment of the subject.

6. The Reynosa manual labor market has changed most noticeably over the last couple of decades in terms of the growth of job opportunities in the *maquiladora* or in–bond export sector. There were 25, 631 jobs in that sector in 1990 and 92,077 were reported in 2005. The 2005 figure put Reynosa in second place, after Ciudad Júarez, in *maquiladora* employment with approximately 19 percent. The traditional sectors of services, agriculture and construction also continue to provide alternative sources of employment for the category of workers who might also consider employment in the brick industry.

It also remains true, however, that brick industry workers generally earn more than workers in other sectors. The only exception currently is in illegal drug–trafficking where young men can earn for one night's work as *narcoburreros* [i.e., essentially as porters who float bundles of narcotics across the river] what they could earn for a week's work in the brick industry. It appears undeniable that the labor pool of able–bodied, young men in the Reynosa labor market is distorted by the persistent demand for *narcoburreros*.

In early 2008 the Reynosos hired four new workers who were recruited from a Nahuatl–speaking region of Veracruz through connections made by their accountant who comes from that same region. They now provide daily van service to the Los Longoria plant from the Reynosa neighborhood where these new employees reside.

Appendix 1. Photographs

1

2

3

4

5

6

7

8

9

10

11

12

13

14

15

16

17

18

19

20

21

22

23

24

25

26

27

28

29

30

31

32

33

34

35

36

37

38

39

40

41

42

43

44

45

46

47

48

49

50

51

52

53

54

55

56

57

58

59

60

61

62

63

64

65

66

67

68

69

70

71

72

73

74

75

76

77

78

79

80

81

82

83

84

85

86

87

88

89

90

91

92

93

94

95

96

97

98

99

100

101

102

103

104

105

106

107

108

109

110

111

112

Appendix 2. Method in the *Border Brick Industry Project* Revisited

The *Border Brick Industry Project* presented a much different set of challenges than any of my previous projects in Mexico (e.g., Cook 1982 and 1984; Cook and Binford 1990). The people and places embraced by the project's scope were considerably more diverse and dispersed and, since they were located on an international border, they were also binational.

One practical implication of cross–border research for those conducting it is the need to negotiate, often on a daily basis, border crossings. Given the high daily volume of pedestrian and vehicular traffic crossing bridges between large cities like Reynosa in Mexico and densely populated areas like the Rio Grande valley in Texas, this can be time–consuming and, at times, disconcerting due to security checks and the need to present credentials. This problem is exacerbated in times of crisis when security–related surveillance is intensified (as, for example, in the aftermath of the 9/11 terrorist attack). Given the riverine locations of many of Reynosa's *ladrilleras*, especially in the Los Longorias *ejido*, and the fact that narcotics traffickers use these as night–time pass–through zones for transporting narcotics across the river, there is often heavy police–military surveillance in these areas that interrupts business as usual for outsiders like *gringo* anthropologists (see Cook 2001a).

In the project's initial phase, sites were covered from El Paso/Ciudad Juárez to Laredo/Nuevo Laredo to Brownsville/Matamoros; in its middle phase these were reduced to the corridor from Roma/ Ciudad Miguel Alemán to Progreso/Nuevo Progreso; and in its final phase to the sprawling municipality of Reynosa.

My first problem as principal investigator was to decide how to approach this spatially diverse array of sites anthropologically in a one–anthropologist project on a limited budget. I realized that it would have been impracticable to include El Paso/Ciudad Juárez in such a study simply because the area's large size and relative remoteness would have required long–term residence and, in fact, would have become an entire project on its own. Moreover, my preliminary research revealed that the handmade brick industry in the Ciudad Juárez/Casas Grandes area in the border region of the state of Chihuahua, unlike the industry of the riverine zone of the state of Tamaulipas, was not export–oriented and produced almost exclusively for local urban construction markets. It shared this local domestic market orientation with brickmaking in Mexican interior states like Oaxaca (Cook 1984), and in other Mexican border states like Coahuila and Baja California. For example, although the handmade brick industries in the periurban areas of Oaxaca City and, especially, Ciudad Juárez, have more permanent plants than Mexicali's industry (Wilson 2005) and, at least in the case of Ciudad Juárez, more brickmakers, Mexicali seems to have more in common with them than it does with the industry in riverine Tamaulipas. A higher degree of capitalist development in the Tamaulipas industry during the second half of the twentieth century set it apart, and reflected its long and intense involvement in a binational economy where production for export to the Texas construction market outweighed production for the domestic construction market.[1]

Within the lower Rio Grande/Rio Bravo corridor (Sánchez, ed., 1994) and riverine Tamaulipas, the municipality of Reynosa became the logical choice as a focal point of research in my project for several reasons: 1) it had the largest number of handmade brick plants in operation; 2) it was directly across the river from the Rio Grande Valley, where my preliminary research suggested, there had been extensive brick production historically; and 3) it was the central place in the brick industry belt in riverine Tamaulipas ranging from the cities of Miguel Alemán, Camargo, and Díaz Ordáz along riverine highway No. 2 to the west and to Ciudad Rio Bravo (and Nuevo Progreso) to the east.[2]

As it turned out, however, the process of site selection to define the universe for ethnographic data collection did not resolve other research problems arising from 'diversity and dispersion' of the industry. It became clear that lower border brickmaking had a binational history that reached into key off–border areas of Texas that would merit special attention. I also realized that, as an academic specialist in Mexican studies and economic anthropology with a special interest in small industry, I had stumbled onto a veritable cornucopia of research possibilities. That cornucopia, however, was entangled in a formidable web of geographical, sociocultural, political, and economic relationships. There were enough research possibilities there to justify several studies.

For better or for worse, I decided that the entire saga of the border brick industry, with all of its complex sociocultural, economic and political entanglements and implications, had to be addressed in one project. In a frame of mind attuned to the conditions of post modernity (Harvey 1990; cf. Cook 1998:283 et. passim), I plunged into the research following a flexible plan that enabled me to move between sites on both sides of the river to visit a brick plant, an office, an archive, and so on according to a shifting agenda—driven mostly by field interviews. Once I learned my way around the lower border region, it was possible to visit different research sites for data collection on one or both sides of the river on the same day from my home base in Weslaco. For example, site visits could be made, and interviews conducted, in one or more of Reynosa's *colonias* [settlements] on the same day, and even in one of these locations and another town like Rio Bravo or Díaz Ordáz. Of course, on many occasions entire days, or clusters of days, were spent at one site whether it was a brick plant or a *colonia* in Mexico, or a county deed records or other archive in Texas.

The research agenda sometimes led me to off–border interviews in large cities like Corpus Christi, Houston, San Antonio, Dallas–Fort Worth, and Austin and smaller ones like Henderson, Conroe, and Tomball. Either on or off the border, I interviewed executives in corporate offices and boardrooms, salesmen in retail showrooms, dealers in distributorships, managers, foremen and workers at plant sites, masons, builders and architects at construction sites or in their offices, farmers at an agricultural cooperative conference room, businessmen in coffee shops or restaurants, and truckers in the cabs of their tractor–trailers. I even interviewed one key informant in the reception area of his daughter's massage parlor in Arlington, Texas!

Finally, I should add that my experience over the years as a researcher in Mexican border sites like Reynosa has been that collecting data from observation of and conversations with living subjects is infinitely more rewarding in cost–benefit terms than searching for documentary data about living or dead populations, wherever it may be found (or not). Admittedly, archival work is more rewarding on the Texas side of the border where the archival/documentary infrastructure is more developed and accessible. But, even there, many records pertaining to the Valley's economic history are either gone or elusive. Something of the frustration and difficulty inherent in any scholarly effort to reconstruct the past was succinctly and pragmatically expressed by one of my informants, Brigitte Waaser Weiske (June 1, 1993) of Madero, Texas: "The trouble with history is that people become interested in it too late after those that were involved with it are passed on."

It was one thing to conduct the research, sometimes at a frenetic pace that was exhausting, if rewarding, and quite another to write up the results. The latter task could not be undertaken until well over 200 interviews were transcribed, read, and analyzed together with substantial amounts of quantitative and documentary data. I developed the concept of "Mexican brick culture" (Cook 1998: Ch.2) around which to organize the book, and my plan was to tell the complete story of the border brick industry in a balanced way without getting bogged down in too much technical and analytical detail, and without losing sight of the present out of a concern to describe what had occurred in the past. I was also equally determined to demonstrate how the industry had evolved interactively on both sides of the river, especially after brick production ended on the Texas side.

That my 1998 book was imperfect in its attempt to tell the story of Mexican brick culture in the lower Rio Grande/Rio Bravo region is suggested by one reviewer's judgment that it was historically shallow, chronologically vague, and too loosely organized, and that "a chronological organization would have served the material better and made it easier for the reader to comprehend the causal relationships that the author is attempting to explicate" (Taylor 1999:172). This critic was unconvinced by my explanation as to how I was trying to be innovative in relating the expositional portion of my project to the research portion in the spirit of "historical ethnography" or heavily weighting research, analysis, and exposition toward oral testimonies from living informants and toward fieldwork (Cook 1998:12). I attempt to do so again in the present book, although I still make expositional concessions to chronology.

In retrospect, I probably did not express my intentions clearly enough regarding how I wanted the 1998 book to reflect the nature of the industry itself, and of the project that produced my knowledge about it. Something along the lines of what I am writing now was called for then. I reject the proposition that knowledge must be presented chronologically because that is the way reality presumably occurs and must be interpreted—one event after another. Must 'reality' be interpreted and presented through such a lens? Does not such an imperative imply that 'reality' evolves chronologically according to some irresistible force, or that somehow 'before' and 'after' translate into 'cause' and 'effect'? My inadequate response to these rhetorical questions is that I have been sufficiently influenced by the postmodern current of thought to be skeptical of chronological thinking and exposition, and their reinforcement of the notion that definitive explanation and understanding of complex processes emerge from their application.

Chronologies have their place but they also make the explanation of complex processes almost too pat, too predictable, too familiar. Chronological thinking seems to pre–empt the possibility of appreciating, for example, how the knowledge of brickmaking contained in the memory bank of a nonagenarian informant elicited in a 1993 interview could have enabled him to make brick in a plant that same year, if he had been physically able to do so. Or of appreciating the fact that deceased border brickmakers from past centuries, if resurrectable from their graves, could successfully practice their craft at a Reynosa brick plant in 2008 just as they had done somewhere in the lower Rio Grande/Rio Bravo region when they were alive. Moreover, if economic and cultural processes evolved according to some chronological law of motion, how to explain the persistence of low–tech artisanal production, supposedly long–ago made obsolete by the inexorable march of technology and economies of scale, in the epoch of globalized high–tech industrialization and mass–market consumerism? In short, chronologies are as limited in their time perspective as are chronometers, they can give you a measurement to locate something according to a particular standard of timekeeping but cannot say with certainty anything else regarding its location there, much less account for its location according to other standards of understanding the meandering streams of time.

Within the framework of my original proposal (Cook 1998:279–286), research in the *Border Brick Industry Project* was mostly dictated by where informants' memory banks led me. They taught me about border brickmaking and its wider connections, and they provided the chronology. I did attempt to verify information elicited from individual informants, including dates, by manually cross–tabulating information from several informants and, when feasible, cross–checking with information from archival or other documentary sources. Like the results of any sampling process, the sampler has to take into account the range of opinions or values and then proceed to find the central tendencies. After being examined and evaluated in terms of knowledge accumulated from other sources, the results of such a process produce explanations and understandings that often transcend chronology.

In drawing his or her own conclusions about these explanations and understandings, the reader should keep in mind that the handmade brick industry is characterized by a paucity of written records. Much of the business transacted over the years in the industry, especially by small–scale *ladrilleras*, has no paper trail. Most of the larger record–keeping plants went out of business in the 1980s (or long before), and their records disappeared with them. This means that the history of the brick industry, not retrievable from the memory banks of living subjects, is very elusive indeed and can only be partially reconstructed, even through sustained, intensive combined archaeological and ethno historical research projects (e.g., Boyd et al. 1994). In short, the only way to avoid writing a shallow history of such a topic is to appreciate the degree to which the past and present are complexly intertwined and resonate with each other within Mexico's commodity cultures, and with the human relationships and lives at their core (Cook 2004). This is a process in which knowledge acquired from living subjects and observations in the present is the indispensable source for understanding the past.

Notes

1. According to Wilson in 2003 Mexicali had approximately 100 brickmaker families (2005:161). I do not have recent estimates for either Oaxaca or Reynosa of the total brickmaker population. One survey of the Reynosa industry in the 1960s estimated that 7,000 men, women and children were dependent for their livelihood on brickmaking. Many plants in the 1960s and 1970s had 100 or more employees. Today, *Ladrillera Reynosa* and *Ladrillera Reynoso* combined have around fifty permanent employees. I doubt that all of the smaller plants still active in Reynosa employ that number of workers, even on an irregular basis. Two brief visits to the brickmaking zones of Ciudad Juárez in the 1990s left me with the impression that it had a substantial population of brickmakers but I have no reliable estimate of its size. I did, however, observe women molding brick there, and also the use of alternative fuels for kiln–firing to reduce pollution. My visit to Oaxaca City in 2004 verified brickmaking activity in several periurban communities—Santa Lucía del Camino, Santa Cruz Amilpas, San Agustín Yatareni, San Sebastian Tutla, and San Francisco Tutla. Again, I have no idea how many households are engaged in brickmaking in these communities but the total probably hovers around the Mexicali number.

2. As one source concisely reminds us (HARC 2008:3): "The local reference to 'The Valley' is misleading; the area is actually a river delta, surrounded by semi–arid Tamaulipan brush land."

Bibliography

Abernethy, Francis Edward, ed. *Built in Texas*. Waco:E–Heart Press, 1979.

Acuña, Rodolfo. *Occupied America:A History of Chicanos*. 2nd edition. New York: Harper and Row, 1981.

Adams, John A., Jr. *Conflict and Commerce on the Rio Grande: Laredo, 1755–1955*. College Station: Texas A & M University Press, 2008.

Allhands, J.L. *Gringo Builders*. Joplin, Mo. and Dallas, Tx: privately printed, 1931.

Barry, Tom, ed.. *Mexico: A Country Guide*. Albuquerque, NM: The Inter–Hemispheric Education Resource Center, 1992.

Bottomore, Tom., et.al., eds. *A Dictionary of Marxist Thought*. 2nd edition. Oxford: Blackwell Publishers.

Boyd, Douglas K., et.al. *Pharr–Reynosa International Bridge: Continued Archaeological and Historical Research at El Capote Ranch Community, Hidalgo County, Texas*. Reports of Investigations, No. 97. Austin: Prewitt & Associates, 1994.

Boyd, Douglas K. "The Guajardo–Vela Ladrillera: Historic Brickmaking in the Lower Rio Grande Valley." *Heritage*, Vol. 13, No. 1 (Winter):20–24, 1995.

Call, B. Ellsworth. *The Geology of Crowley's Ridge*. Annual Report of the Geological Survey of Arkansas for 1889. Vol. II. Little Rock: Woodruff Printing Company, 1891.

Campbell, James W.P. and Will Pryce. *Brick: A World History*. New York: Thames and Hudson, 2003.

Cook, Scott. *Zapotec Stoneworkers: The Dynamics of Rural Simple Commodity Production in Modern Mexican Capitalism*. Lanham, MD: University Press of America.

______. *Peasant Capitalist Industry: Piecework and Enterprise in Southern Mexican Brickyards*. Lanham, MD: University Press of America, 1984.

______ ."Inflation and Rural Livelihood in a Mexican Province: An Exploratory Analysis." Mexican Studies/Estudios Mexicanos 4 (1):55-78.

———. "B. Traven and the Paradox of Artisanal Production in Capitalism: Traven's Oaxaca Tale in Economic Anthropological Perspective,"*Mexican Studies/Estudios Mexicanos* 11(1):75–111, 1995.

———. *Mexican Brick Culture in the Building of Texas, 1800s–1980s*. College Station: Texas A & M University Press, 1998.

———. "The Mexican Connection in the Southwestern Borderlands: An Introduction," *Journal of the West* Vol. 40, No. 2:5–14, 2001a.

———. 'In the Maquiladoras' Shadows: The Handmade Brick Industry in Transborder Development,' *Journal of the West* Vol. 40, No. 2:32–41, 2001b.

———. *Understanding Commodity Cultures: Explorations in Economic Anthropology with Case Studies from Mexico*. Lanham, MD: Rowman & Littlefield Publishers, Inc. 2004.

———. "Bricks, Aesthetics and the Market." *Texas Architect*, Sept./Oct: 26–27, 114–115, 2005.

———. "Ladrillos, Ladrilleras y Los Ladrilleros: Brick Culture and Architecture in Lower Rio Grande/Rio Bravo Border History." *Detalles* (newsletter of the Lower Rio Grande Valley chapter of the American Institute of Architects) December, pp. 6–7, 2006a.

———. "Ruins of Brick Culture Strewn Along Lower Rio Grande." *Texas Architect*, January/February: 25–26, 2006b.

———. "Historical Brickmaking in the Two Laredos." *Journal of South Texas*, Vol.23, No.1 (Spring), 2010.

Dillon, David. *The Architecture of O'Neil Ford, Celebrating Place*. Austin: University of Texas Press, 1999.

George, Mary K.H. *O'Neil Ford, Architect*. College Station: Texas A & M University Press, 1992.

Gurcke, Karl. *Bricks and Brickmaking: A Handbook for Historical Archaeology*. Moscow, Idaho: The University of Idaho Press, 1987.

Hansen, Lawrence Douglas Taylor. "The Origins of the Maquila Industry in Mexico." *Comercio Exterior*, Vol. 53, No. 11, November: 1–16, 2003.

Harvey, David. *The Condition of Postmodernity: An Enquiry into the Origins of Cultural Change*. Cambridge, Mass.:Blackwell, 1990.

Herrera Pérez, Octavio. *Monografía de Reynosa. Investigación histórica e iconográfica*. Cd. Victoria, Tamaulipas:Instituto Tamaulipeco de Cultura. 1989.

Horgan, Paul. *Great River: The Rio Grande in North American History*. Hanover and London: University Press of New England, 1984.

Houston Advanced Research Center (HARC). *Water in the Lower Rio Grande/Rio Bravo Basin*. www.harc.edu/Projects/Archive/RioGrandeBravo, 2008.

Isbell, Frances W. ed. *Hidalgo County Ranch Histories*. Edinburg, Texas: Hidalgo County Historical Society and Hidalgo County Historical Commission, 1994.

Johnson, Marjorie, ed. *Valley Proud History Cookbook*. Weslaco, TX: Rio Grande Valley Chamber of Commerce, 1991.

Kearney, Michael. *Reconceptualizing the Peasantry: Anthropology in Global Perspective*. Boulder: Westview Press, 1996.

Kelly, Annie and Tim Street–Porter. *Casa Mexicana: The Architecture, Design, and Style of Mexico*. New York: Stewart, Tabori and Chang, Inc., 1989.

Martínez, Oscar J. *Border People: Life and Society in the U.S.–Mexico Borderlands*. Tucson: University of Arizona Press, 1994.

McBride, John. *Vanishing Bracero: Valley Revolution*. San Antonio: Naylor Company, 1963.

Montejano, David. *Anglos and Mexicans in the Making of Texas,1836–1986*. Austin: University of Texas Press, 1987.

Morales, Rebecca and Jesus Tamayo–Sánchez, "Urbanization and Development of the United States–Mexico Border", in *Changing Boundaries in the Americas: New Perspectives on the U.S.-Mexican, Central American, and South American Borders*, Lawrence A. Herzog, ed., pp. 49–68, 1992.

Paredes, Américo. "The Problem of Identity in a Changing Culture: Popular Expressions of Culture Conflict along the Lower Rio Grande Border." In *Folklore and Culture on the Texas-Mexican Border* by Américo Paredes, edited by Richard Baumann, pp. 19-47. Austin: University of Texas Press, 1993.

Peña, Manuel. *The Texas–Mexican Conjunto: History of a Working–Class Music*. Austin: University of Texas Press, 1985.

———. 1999. *Musica Tejana: the Cultural Economy of Artistic Transformation*. College Station: Texas A & M University Press.

Sánchez, Mario L., ed. *A Shared Experience: The History, Architecture and Historic Designations of the Lower Rio Grande Heritage Corridor*. 2nd edition. Austin: Los Caminos del Rio Heritage Project and the Texas Historical Commission, 1994.

Searle, Alfred B. *Modern Brickmaking*. Fourth Edition Revised and Enlarged. London: Ernest Benn Limited, 1956.

Steinbomer, Robert. "The Texas Brick: A Brief History of the Industry and Its Product," *Texas Architect* January/February:44–47, 1981.

Taylor, Lonn. Review of Scott Cook, *Mexican Brick Culture in the Building of Texas, 1800s–1980s*. *Winterhur Portfolio* Vol. 34, No. 2/3 (Summer–Autumn). 168–172, 1999.

Traven, B. *The Night Visitor and Other Stories*. Chicago: Elephant Paperbacks, Ivan R. Dee Publisher, 1993.

Walsh, Casey. "Aguas Broncas: The Regional Political Ecology of Water Conflict in the Mexico–U.S. Borderlands." *Journal of Political Ecology* 11:45–60, 2004.

———. *Building the Borderlands: A Transnational History Of Irrigated Cotton Along the Mexico–Texas Border*. College Station: Texas A & M University Press, 2008.

Wilson, Tamar Diana. *Subsidizing Capitalism: Brickmakers on the U.S.–Mexican Border*. Albany: SUNY Press, 2005.

Zamora, Emilio. *The World of the Mexican Worker in Texas*. College Station: Texas A & M University Press, 1993.

Index

About the Author

After completing undergraduate studies at the University of Texas and American University (B.A. economics, 1959), **Scott Cook** did graduate work at the University of Wisconsin (economics), the University of Puerto Rico (Graduate Diploma in Caribbean Studies and Social Sciences, 1962), and the University of Pittsburgh (PhD anthropology, 1968). He taught anthropology at Michigan State University (1968–1971), and at the University of Connecticut (1971–1997). At the University of Connecticut, he also directed the Center for Latin American and Caribbean Studies for five years (1987–1992) and the Puerto Rican and Latino Studies Institute for four years (1996–2000). During 2002–2004 he was a Visiting Scholar at the Long Institute of Latin American Studies at The University of Texas–Austin, and in the spring of 2004 he served as an Adjunct Lecturer at the University of Texas–San Antonio.

His most recent book is *Understanding Commodity Cultures: Explorations in Economic Anthropology with Case Studies from Mexico* published in 2004 by Rowman & Littlefield. In 2001 he edited a series of essays entitled *The Mexican Connection in the Southwestern Borderlands—Trends and Prospects in the Journal of the West* (Vol. 40, No. 2). Four earlier books and many journal articles dating back to the 1960s address various themes in economic anthropology, peasant–artisan studies, Mexican studies, and border studies, and draw upon extensive fieldwork in Oaxaca, Mexico and the Texas–Mexico border. Over the last decade, he has lectured and published in Texas and the lower Rio Grande corridor on the topic of the handmade brick industry.

Scott Cook has served on the editorial boards of *Dialectical Anthropology*, the *Journal of Peasant Studies*, the *Journal of the West*, and *Mexican Studies/Estudios Mexicanos*. He has lived in San Marcos, Texas since leaving the University of Connecticut as Professor Emeritus in 2000, and can be reached via e–mail (hscook@grandecom.net).

www.ingramcontent.com/pod-product-compliance
Lightning Source LLC
Chambersburg PA
CBHW051253200926
51612CB00001B/32
9780739147979